DECODING EARTH'S HIDDEN SECRETS

DECODING EARTH'S HIDDEN SECRETS

CURTIS E. WOOD

TATE PUBLISHING
AND ENTERPRISES, LLC

Published by Tate Publishing & Enterprises, LLC
127 E. Trade Center Terrace | Mustang, Oklahoma 73064 USA
1.888.361.9473 | www.tatepublishing.com

Tate Publishing is committed to excellence in the publishing industry. The company reflects the philosophy established by the founders, based on Psalm 68:11,
"The Lord gave the word and great was the company of those who published it."

Cover design by Samson Lim
Interior design by Jomar Ouano

Published in the United States of America

ISB0N: 978-1-68142-723-2
Science / Global Warming & Climate Change
15.05.19

To the world and all the children here today and to those who are yet to come. This is for those who have no vote, no voice, and no say in their future. My three grandchildren—Charlee, CJ, and Connor—have all been my primary incentive and driving force that continues me through many sleepless nights of researching information and pouring through databases. I hope that the knowledge contained within this book helps guide them and helps prevent mankind from the continual cycle of self-destruction he repeatedly creates. If humanity, as a whole, can learn from the past and apply it in the future, perhaps we will be able to evolve to a new and enlightened level. The alternative is to continually repeat past concepts, ideas, and beliefs that have only resulted in massive destruction and suffering. Knowledge and understanding is the key to our success and failure—both today and in the future.

To both my mother and father. I required a team effort on both their parts, and for this reason, it is difficult to separate them. My mother just passed away and was truly the most influential person in my life. It was clear from a young age that we shared a passion for nature. She always had a way of encouraging me and lifting me up when I felt hopeless. I know that it was her patience and countless days and nights of discussions and listening to me that now brings all my research to you.

When I think of the former methods of research by going to the library and trudging through countless index cards, books, and articles, I realize that such expedient and thorough research was never possible until today—the information age. The technicians that keep the internet running, the people who assembled my computer, the salesperson who sold me the computer, the truck driver who delivered the computer, an endless list until it eventually reaches you: the reader. We live in a society today that is intertwined and interconnected worldwide to an extent that until today could only be imagined. So to every one of you, I thank you.

CONTENTS

INTRODUCTION

When all of this information was assembled, I had to weigh the possible side effects that could arise from releasing this information, so I went to a number of individuals I respected and posed the following question to them.

"Now that you know this information and releasing it could potentially create a state of chaos, my question is…do the people have the right to know?" Universally, one after another, they responded, "Yes, the people have the right to know." So this knowledge is my gift to be shared with the world and future generations.

Sitting down and writing this book has become a very difficult task for me. I was talking with some friends the other day and one of them asked me when I planned to finish my book. It has been in progress for some time, and I told him that I was having difficulty discontinuing my research in order to begin assembling it all. He then looked at me and asked me, "When will you be done with your research?" I realized at that moment that I would never stop researching and exploring questions as long as I live, but the time has come to assemble all my notes into this book.

Additionally, I came to realize it would be beneficial to have other people involved in this research and to pass on this knowledge and understanding. I hope this will bring many of the scientific fields together in their research and data. By using each other's knowledge and understanding, they may be able to improve methods of monitoring, impose warnings as necessary, and in the end, save lives.

Although I am writing this book for the general public today, the primary purpose of this book is for the children of our future—the ones who will be left with a world that we are destroying. This book is meant to act as a guide and manual for living upon this planet and understanding where we went wrong so we can prevent making the same mistakes repeatedly. Perhaps their world can become the world that we have dreamed of, but this can only be accomplished if our Earth is completely and fully understood.

After completing school and having children of my own, my mother began to open up to me about my early years while I was in school. She felt that this would help me with some of the issues that I was experiencing with my own children. She told me that the first day home from kindergarten, she had asked me, "How was school?" She then began to chuckle and while giggling told me that I looked at her and said, "Fine, but do I have to go back?" I continued to carry this attitude with me throughout my life. I never enjoyed elementary school and being confined to a desk most of the day; it frequently seemed more like torture than education.

We use to have to take the IOWA tests in school that were drawn out and time consuming. I hated them. Bored with the test within the first thirty minutes, I would give up and start filling in the dots, mixing them up and trying to guess a pattern (that apparently was wrong all the time). My mother told me many years later that she would have to go to the school each year and explain to the teachers. She told me she never worried about me and reading because I was always busy looking up football and baseball statistics, reading short stories and articles, and spent hours researching topics such as nature and wildlife in various encyclopedias.

Often, we would go down to the local pond, lake, or creek and pretend we were explorers like Lewis and Clark. It was during these times when I would dream of living back in those days and being able to see the land before man came and cut it all down.

I was very curious as a child, and this part of me has remained unchanged. Nothing went into the garbage without dissecting it first to see how it worked, or more often than not, what's inside that makes it work?

Entering junior high was a huge relief, and although I still hated sitting at my desk, I was at least able to get up and move around every hour. At high school reunions, we have discussed this transition and there is universal agreement: junior high was a great relief for all of us boys for this reason. Shop and art classes were the most enjoyed because the desks were rarely used, and our imaginations and creativity were encouraged. I inherited my mother's love for nature, and even today, my art acts as an escape for me. I sit back and think of a location in my mind where I would like to be, and the feelings and emotions that are created inspire an image onto the canvas.

I had never heard of attention deficit disorder (ADD) until my son was first diagnosed with it. When they told me he had ADD, I quickly responded, "Of course, he's my son." The words seemed to describe me. Today, they would have placed me and most of the boys back then on medications in order to keep us confined to a desk all day. As I told my son years later: "We do not have ADD, we have SADD." He looked at me baffled, and I told him that most men have this disorder that I call "Selective Attention Deficit Disorder." He had just finished a portrait of a friend and spent countless hours indulged in this drawing when I asked him, "How can you have difficulty with attention, and yet be so fixated and focused on a drawing for so many days?" This is when I explained to him that his diagnosis was not a disorder, but a gift that others see as a disorder and is typical and normal in many boys. Boys and men have great difficulty remaining focused on areas that they have no interest in and frequently obsess themselves exceedingly in areas that do interest them. Today, I use a simple rule for obsessions: am I in control of my obsessions or are my obsessions in control of me? The key I have learned in life is balance and to not allow any obsession to interfere with my personal responsibility, after all, obsessions are a part of being human.

I grew up in a Minneapolis suburb with a father in the Sheet Metal Union, so heating and air-conditioning has always been a part of my life. In elementary school, I thought everyone knew what an "R factor" was because it was kitchen table talk when I was young. The R factor is the value of insulation that is, for example, placed in the walls and ceilings of your home that resists the flow of energy in the form of heat. I became more curious about his work in my teen years and became a laborer working for the Sheet Metal Union on various jobs during the summers in high school. During these same years, my interests quickly turned to cars, especially the day I got my license and a 1967 Mustang. After high school, I entered a vocational training program for auto mechanics and continued to advance my education. I was born with a racing spirit in my blood and loved to go fast—so fast that soon my speeding tickets were catching up with me. In order to avoid further complications, I decided to join the military.

I again furthered my education when I became a motor transport mechanic in the United States Marine Corps. When I arrived in Okinawa, Japan, the lieutenant informed me that she was going to have me working in the office with her and "the top." The top is the chief enlisted man who was a master sergeant. When I told her I didn't want to work in the office, she told me that my test scores are borderline genius and she needed me there. I then turned to her and said, "With all due respect, ma'am, I don't care what my tests say. I will go crazy sitting behind a desk all day, and I'll end up driving all of you out of here." She just sat behind her desk and looked at me with a very curious look because I was choosing a far more difficult job and the heat in a shop over an air-conditioned office. She must have seen something in my face that told her I was being truthful, so she decided to place me out on the shop floor, and I was soon in charge of the inspection rack as the company's motor pool inspector. All vehicles coming in and out of the shop, including all scheduled maintenance, were my responsibility to verify the safety and repairs of every vehicle before leaving the shop.

When I was honorably discharged, I decided to alter my training and wanted a challenge, so I chose computer technology. This was a two-year self-paced course that I completed in 18 months. Electronics became an easy crossover for me because many of the theories and formulas mirrored heating. Both use energy, resistance, and directional flow with many of the same calculations, only different variables. I then found employment working in the access control industry for three years before starting my own business, and I haven't punched a time clock since

1988. I started my own business in 1988, installing and servicing security systems including access control, video surveillance, fire alarm, and a vast array of low voltage systems. I have had my highs and lows throughout the years with my business, and today I am also an entrepreneur and an author. I own my own business where I work during the day, but on nights and weekends, I indulge in my research efforts and statistical analysis to find answers to problems and questions that come to mind.

I have never officially taken a college course, but this has not kept me out of college classrooms. I have given numerous seminars at the local college and community groups on various topics, including climate change, and although this is my first published book, I am not new at publishing either. I have written several articles that have been published over the years in the local newspaper on various topics. For more than a decade, my research has involved a very wide spectrum of topics including the Israel–Palestine conflict, government taxation, income distribution, oil and gas, and most recently, weather. Much of it, in the end, has been necessary and has been compiled in some way into this book.

Family members and close friends frequently tell me, "You think too much." It has taken most of my life to understand that something in my mind differs from many people. My mind can both see and follow trends, patterns, and alterations, especially energy flow that easily helps me with troubleshooting. This ability is what makes me a master troubleshooter today in electronics—finding anomalies such as ground faults, induction, and ground potential differentiations—looking for and finding things that can't be seen. For me, electricity is simple, and as an instructor once told me, it's lazy and just wants to find the easiest path to ground. My mind visualizes electron flow like heat traveling in a unidirectional path, making it easy to find leaks, distortions, degraded signals, or inductions coming in from another source. Heat is the same way and wants to find the easiest path to a cooler, calmer environment, so I felt that my mind and years of troubleshooting made me well suited for this task. All heat wants to do is equalize, and this state is known as thermal equilibrium.

Over the past few years, I was made aware of an arachnoid cyst residing behind my right eye. This was diagnosed via magnetic resonance imaging (MRI) after complaints of pressure sensations being felt in that location. This is one abnormality of my brain that does appear to be different from most people, even those with cysts because of the location. Although rare, it's not life threatening and I consider it a gift. I believe that this has contributed in some way to my thought process and has enhanced my creativity.

My life experience and understanding of heating and cooling made this research a challenge for me, and by far the greatest machine to decode due to the complicated and intertwined nature of its operation. If the planet was heating, which has seemed evident, then I felt I could locate the source and cause if I possessed a strong enough desire. The love for my children and their children has, and continues to drive me today, and will to the end.

When I decided to take on the planet and the global warming project, I realized that no manuals existed regarding the heating and cooling of the planet. Science was able to provide a vast amount of data and research, but a definition of the way everything works together along with the cycles was only found in bits and pieces. Examples such as water, tectonic plates, volcanoes, carbon dioxide, or the atmosphere were readily available, but independently. In my history, whenever I come across a machine and no manual is available, I begin by creating my own. Utilizing previous experience and known physics, I begin to reverse engineer the machine and then move forward with troubleshooting and repairs. In this situation, the machine or engine is the planet that is powered from the sun and the core along with the gravitational forces placed upon it.

Understanding the operation and the diagnostics of a machine are two totally different fields of science. Troubleshooting requires a thorough understanding of all the parts associated within a machine and how they interact along with complete and thorough understanding of the physics at work. It is what I term the "if/then theory of operation." This is the understanding that if a certain situation occurs, like a freezer door being left open, then the food will thaw. If it remains warm for too long, it can cause the food to begin to rot. If you then eat the food, you could get sick from food poisoning. One problem can cause multiple effects, and something as complicated as the earth, I am sure, has many of these scenarios.

This is not a field that is taught in any textbook or classroom, and this requires time, patience, and years of field experience. Using common sense and logic, I will let the research and discoveries found along the way stand on their own merit and allow you, the reader, to decide. To begin this quest, we need to go back and start at the beginning with one simple question that opened a floodgate of others.

IS THE PLANET WARMING?

Temperature

Science of the Earth is the discovery, defining, and explaining the basic principles that govern the earth. This has become a way for man to understand the world, like the freezing point of water being 32 degrees Fahrenheit, or zero degrees Celsius. This is just man's reference to a known alteration in the state of water. The Earth created all these properties and man, through science, has defined it and at times, redefined it.

The first step of troubleshooting is to first fully understand what you're working with. Look to the basics and principles, and then understand what is often termed the theory of operation. When the theory of operation is not present, like an outdated machine where a manual no longer exists, you have to draw from past experiences and look back to the basics and principles to draw your own manual of understanding. The Earth's heating and cooling manual, the principles of this operation, have not been fully explored so they are not available. As a result, it was necessary to begin by taking this approach for my research: by understanding all cycles and principles involved in the heating and cooling of our planet in order to understand the way it operates, but more importantly, what we can expect so we can better prepare ourselves and make changes as needed. I knew there needed to be answers to past anomalies such as the great Dust Bowl. The heat of this era was never fully explained, but what was obvious was the dust that gave the era its name. Later, we experienced cooling and were told we were entering the next ice age. Now, with recent global warming accelerating at an unprecedented rate, I was compelled to use my years of troubleshooting abilities and techniques, along with the knowledge I have obtained over the years, and apply them to the Earth. So, the question that came to my mind was, "Is the current alterations we are experiencing natural, or has it been stimulated by man?"

Science is divided into many fields, and many of these fields explore specific areas of heating and cooling of the planet, but they do not cross over into one another's areas of research. A brain and heart surgeon both work with the human body, but they do not cross paths of specialty and remain focused on their specific area of research. The same applies to the aspects of the study of Earth science. I began compiling data from multiple sources and used this information to cross-examine other data. Without the use of current technology, I would have never been able to compile the information and the numbers over the past several years. The amount of information acquired over such a brief period of time would have previously taken several lifetimes to acquire.

There are many areas that need to be researched that involve a variety of aspects within the scientific community. None of the data I used was my own, but years of research, study, and documentation along with the hard work of so many people both within and outside of the scientific community. It's an accumulation of their data and compiling it in unique ways that has allowed me to uncover these results and findings. So I must thank the millions of researchers, scientists, students, and volunteers that have allowed me to gather and present this data.

I had heard about global warming and climate change, but never really gave it much thought until I found out I was going to be a grandfather during the 2010–2011 holiday season. Age, time, life-altering changes, and now becoming a grandfather prompted me to think about the future and presented these questions: What kind of life was this child going to have? What kind of world would we leave her and the others? I had so many questions with no real answers.

When Charlotte was born, my little Charlee, questions about the future of children today became more passionate. I sat down and held her and soon found myself thinking of her future—her first day of school, graduation, marriage, and her children. My life was to make a dramatic shift, and I became obsessed with a deep passionate quest to understand our world and do whatever I could to make this a better world for her and the other children who have no voice today. Now, with the addition of two adorable grandsons, Connor and CJ, my passion has only intensified.

The Quest

I began to question warming a long time ago as our winters began to shift. I recall the first time I made a comment was at our going away party when my friend and I entered the Marine Corps in late December of 1979. The temperatures were so warm that many people were not even wearing jackets, and it felt like spring. Never before was the weather this warm this time of the year, but we truly enjoyed it. Global warming in the northern climates tends to be an event that is received in gratitude rather than despair.

More recently, I was beginning to see possums coming into the area, a strange sight for central Minnesota. I had one alarm system that had motion detectors going off in their yard every night. He had problems in the past with cats, raccoons, and skunks, so he decided to set out some live traps. What he didn't expect was what he saw in the morning—a very upset possum. Within a few months, my parents saw one climb their fence in their backyard. My dad told me it was the most awkward thing he had ever seen. He told me it climbed right up the fence, and then head first down the other side and then continued on without missing a step. Today, they can be seen as far north as Brainerd alongside the roads. I had enjoyed the outdoors my entire life and had never seen one here. When I asked my dad who had spent a great deal of time outdoors when he was young, he told me that he had never seen one in the wild here either.

When I did my research on the possum, I found they do not hibernate so they must look for food throughout the winter. With bare skin on their ears, nose, and tail, they are not found in extreme cold environments. They are attracted to wetlands, which we have plenty of, but our winters keep them farther south. When prolonged cold spells hit, what deer hunters once called a "thinning of the herd," the possums would die off. By looking to nature and the possums' extended range, there are clear signs of some alterations occurring. But still the question remains, is this a cycle as many people state, or are we causing it?

In order to begin to get a visual concept of the amount of heat we are dealing with, I looked around and figured that my car was the one element that would produce more heat than anything I use. I decided to run some figures and brought it over to the local college and talked with some of the instructors. Each car and truck on the road averages 330,000 Btu at an average fuel consumption of 20 miles per gallon of gasoline. This means that one car can easily heat three average size homes in a Minnesota winter. The only difference between the automobile and the home furnace is the automobile is engineered to dissipate as much heat as quickly as it can for power where a furnace is engineered to save and capture this heat. When calculating the heat distributed by our use of gasoline and diesel fuels alone around the world, we found that the amount of energy was equal to the Hiroshima bomb going off at ground level every thirty seconds, forever. Although an astronomical amount of heat, when compared to the sun, it was only a drop in a bucket. I quickly understood that there is no way we can physically alter the heat through the production of it.

As I began to explore global warming, or climate change, whichever may be more politically correct, I decided to turn to my mother and father and discussed this with them. They have many more years here on Earth and first hand account is always the best. Many people indicate that it's only a cycle, while the science community states that warming is caused from an increase in carbon dioxide levels from our use of fossil fuels. Both arguments have credence to their position, and carbon dioxide is a warming agent in our atmosphere, but is it enough to make the changes we are experiencing? I wanted to compare their memories along with mine and attain some idea to the patterns and cycles experienced in the past, and then compare them to today.

My parent's memories brought them back to the late 1930s and '40s, a period of time when the world seemed to be burning up. When I was young, I was being told that the world was heading toward another ice age, while they learned in school that we were heating up. With these variables, it is easy to understand why people would see this as a cycle. It does often seem as if the world is going through cycles, but is it?

As my mother began to tell me about her childhood, she told me that her mother had felt badly because she was born in May of 1931, and that summer was one of the hottest years on record. She said that her mother had told her that it was so uncomfortable that holding her only made both of them more miserable. Her memories back then were spent trying to find ways to be cool during those summers and spending much of the time around the lakes. Air-conditioning was not available, but there were icehouses where many people would gather together on hot days, especially the men.

I was curious about the winter months, how they were when they were young. Both of my parents agreed that it would get extremely cold in the winter, but not like it is today. They felt that it had cooled more when I was young in comparison to their youth, but they indicated that this was more prevalent with the summers. There was one month that we all remember when the high temperatures for the day never went above zero until the last day of the month. After going a whole month, we were all disappointed that it went above zero that last day, at least everyone except my father. He had to work outside on occasion, and during this period of time, he was working outside the entire month, and he reassured me that the winter months are far warmer than those days. My mother chuckled as she recalled that time, and they both agreed the extreme cold was still present back then, often reaching temperatures in excess of 20 degrees below zero.

My mother continued to tell me about the days it would snow and how beautiful and white it was. Then she looked at me and told me, "That was when it was fun to go out and play in the snow. By the next day, the snow would be all black." When I asked her about this, she told me that in those days, everyone used coal to heat their homes and businesses, and it would blanket the skies with soot and darken the snow. I recall when I was young cutting the snow and viewing the layers because air pollution was so bad that you could count the past snow falls by the gray lines. In comparison to her youth, it was much worse. Although layers within the snow can still be seen and is natural as dust and other elements land on it between snowfalls, these layers today are not darkened in soot as they once were.

My mother told me that she was living in Los Angeles for almost a year and had been told that there were hills you could see on a clear day just beyond the city. When I asked her if she ever saw them, she told me, "Once, just once." The news often displays the pollution over Beijing, and this could have easily been any city in the United States while

I was growing up. Air pollution was considered industrial progress and was a part of life and nothing to be concerned about early on, even in my days.

I walked away from our discussion understanding that the summer heat had altered in an oscillating form over the decades, but our winters had clearly seemed to alter. This has become more apparent over the years as we would head out for deer hunting in November. In recent years, we often find ourselves in just a sweatshirt and sometimes a T-shirt while our cold weather clothing remained stored in our closets. I decided to look into this and used this as the starting point for my research. Are we warming?

Minneapolis is located at the 44.98 parallel, meaning it is the halfway point between the equator and the North Pole and has continental climate conditions. Continental climate is an area where the weather system is not directly influenced by any ocean or sea and has greater deviations in temperatures as a result. Due to its location in the middle of the North American continent, this location is an ideal place to run statistical analysis reports. I turned to the climate page at the University of Minnesota and began running statistics.*

In order to observe the extreme cold that we felt had been declining, I decided to add up all the below zero

---

* All data compiled for Minneapolis weather was attained from http://climate.umn.edu/doc/twin_cities/twin_cities.htm.

temperatures for a season and graph them out annually. The deep Arctic cold temperatures that would plummet to more than 20 degrees below zero were common, growing up for both my parents and myself, so this would reveal the Arctic's influence on our land. The term Alberta Clipper was often used to describe these cold blasts. Because this timeline overlaps years, the year plotted is the fall of the previous year including the spring of the year plotted. For example, 2012 indicates all the below zero temperatures from the fall of 2011 through the spring of 2012.

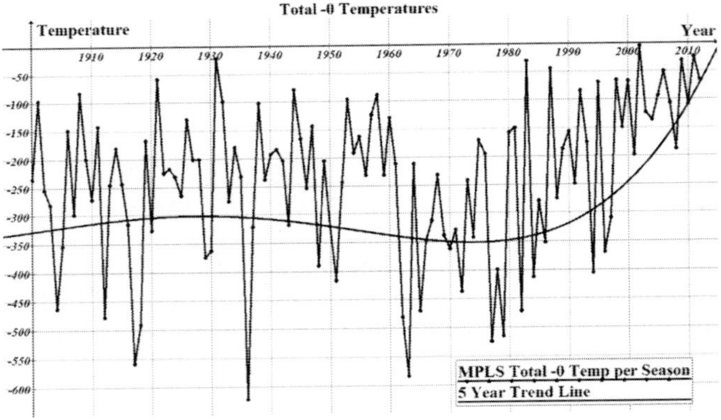

This graph confirmed what I had expected: a warming, cooling, followed by another warming, as well as a significant difference today in comparison to the 1930's era. The average below zero temperatures per season would run approximately –300 to –350 degrees below zero with a significant decline after 1960 until the 1980s, and afterwards we see a continual warming trend. There was a slight rise that preceded the 1930's era that declined but has risen since. The last cold winter season experienced in Minnesota according to this graph was in 1996, and we have been seeing warmer winters since. According to this graph, if the current trend is to continue, Minneapolis will have a fifty-fifty chance of not dropping below zero for an entire winter season by 2015. In 2001–2002 winter season, there were only two days in which the temperatures dropped below zero. Minneapolis is on the edge of this cold Arctic influence, and this graph clearly indicates a receding Arctic just as if you were watching a lake dry up year after year.

This explains the presence of the possum; they are wandering into our area because the temperatures have allowed them to survive our harsh winter environment since 1996. When I was young, the temperatures during many winters would drop to –20 degrees or lower for extended periods of time. When the winters become this cold, only the strongest survive and migratory species are eliminated. In the recent 2013–2014 winter season, there were some extreme cold temperatures that would have some impact on these species, but how much at this time is unknown. Future extreme cold winters will be necessary for some species, especially those fighting invasive insects.

When I was young, I had severe hay fever that made me miserable. In the fall, after it would freeze, my allergies would disappear. I would always look forward to the onset of autumn, and I remember my mom telling me that we needed a hard frost to kill everything, temperatures that would drop to 30 degrees or less. This would occur shortly after my birthday in late September or early October, so I decided to see when we achieved this temperature in the fall and see what it would reveal.

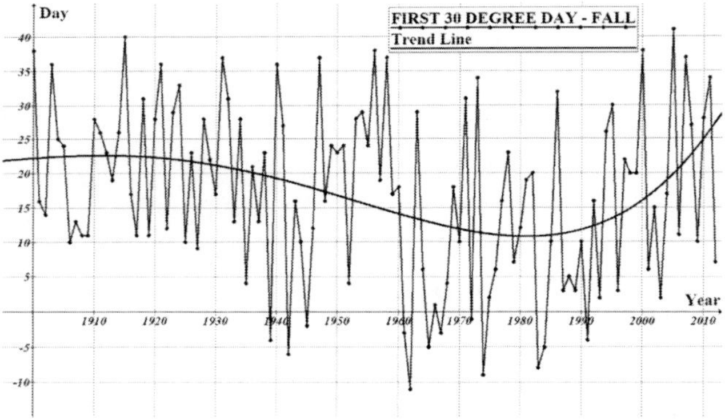

The y-axis represents the day in October it first hit 30 degrees. A negative number indicates September and after 30 is November. The x-axis indicates the year.

This graph indicates the same slight rise, fall, and rise, which explains the changes that have occurred causing an extended growing season and a delayed onset of the winter season. This also explains the deer hunting clothes that are seemingly never worn anymore. I then decided to continue with this research and look to the spring and see if the same temperature curves were taking place.

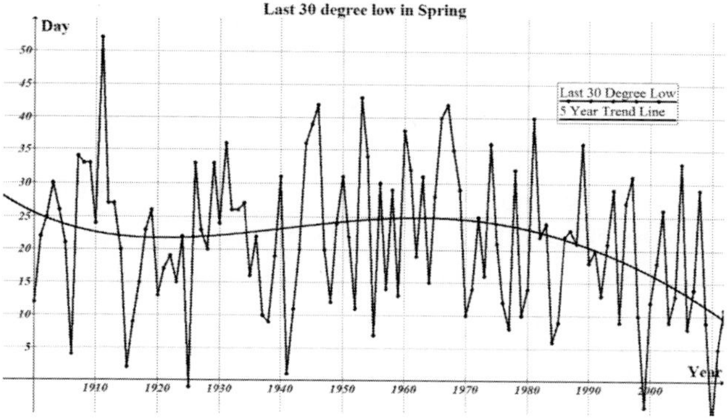

The x-axis represents the year, the y-axis represents the day in April the last day the temperature was 30 degrees. A negative number indicates March, and after 31 is May.

In this graph, we view a reverse image of the other two graphs. The reason for this is the dates we are achieving this temperature is earlier in the season, so warming is indicated by a downward rather than an upward trend. By using these three graphs, I was able to ascertain that the winters are altering and confirmed our memories. The spring is coming earlier and the fall is lengthening, causing a shortened winter season. This is good for the crops by extending the length of the seasons, as long as there is proper distribution of water taking place.

All three of these graphs indicate a universal rise, reduction, and then rise in heat that we are experiencing today. Understanding that the sun and the core of the earth are the only two sources of heat for our planet, I began to think about pollution levels. I then recalled the levels of pollution that were experienced in my day, and then what my mother talked about in the early days. In 1970, the Clean Air Act was enacted, and the Environmental Protection Agency (EPA) was instituted. I could see how reducing the solar radiance upon the surface could result in reducing our temperatures. I compared it to the heat difference felt between areas in the shade in comparison to being in direct exposure of the sun on a hot summer day.

I researched this and found that this dip in temperatures is defined today by science as a period of "global dimming." The obscuring of solar energy upon our planet with pollutants in the atmosphere results in a period of cooling. If this initiated a cooling effect upon our planet, then it would only be logical to draw the conclusion that by cleaning the atmosphere, over time, the heat would return. If this was the cause and effect, then we would actually need to better understand the 1930's Dust Bowl because this would indicate a return to this period of time. Whatever factors were driving our continent's weather in those days is only returning once again. It was clear that there were still many questions that needed to be answered.

Understanding this, we can conclude that the solar radiance upon the surface had a greater impact on the Earth's temperatures than the carbon dioxide levels that were constantly rising during this time. So even though carbon dioxide is a warming mechanism, solar radiance, or a reduction in it, has a greater effect on our temperature and weather than carbon dioxide levels. This does not discount carbon dioxide as a warming mechanism; this indicates that at the levels at these times, the impact was minor in comparison to the alteration of the incoming sunlight.

Our vast use of fossil fuels began in 1908 when Henry Ford began mass production of the Model T, and then expanded rapidly during World War I. When comparing society's output today, the early days were rather insignificant prior to the current time. Since carbon dioxide was beginning to rise in the 1800s before any large scale use of fossil fuels began, it is necessary to explain this as well. Temperatures rising by the late 1800s indicate there must be another cause. This does not discount carbon dioxide as a cause of this warming effect in the atmosphere, but indicates that this early rise in carbon dioxide levels was a by-product of another source.

It is clear by both the factual data and firsthand accounts that we are warming again and on a greater scale than the 1930's era. I felt my understanding of thermodynamics and heat flow would be able to lead me to the source. If I could find this, I figured the cause would reveal itself.

The cause of the great Dust Bowl is classified as improper farming techniques. World War II began shortly thereafter, the planet began to cool, and books were closed on this subject. This research picks up where these scientists left off.

EARTH'S BASICS

The Earth

I decided to first look at our planet from the basic fundamentals of known physics. Because the planet works on cycles, we must first identify these cycles, how they work, and then we can assess how these cycles could have altered. Altering any part of these cycles in any manner will affect the planet's natural heating and cooling processes, resulting in altered weather patterns. Understanding these cycles, the physics involved, how they interact, along with historical alterations and events are researched and examined.

The Earth has a total landmass of 57.5 million square miles, and the landmasses are disproportionate between the northern and southern hemispheres. In the northern hemisphere, there is a 1:1.5 landmass-to-water ratio compared to a 1:4 land ratio in the southern hemisphere. In the northern region, there is 38.7 million square miles of land compared to the southern hemisphere having 18.8 million square miles. When we subtract Antarctica that is covered in ice, we find there is 13.4 million square miles of open land in the southern hemisphere, three times less than the north. In the southern hemisphere, we also find vast amounts of harsh environments such as the Atacama Desert of Chili and the Australian outback consuming 2.54 million square miles between them, or 19% of this available land. This is the result of the southern hemisphere's summer, occurring while the planet is closest to the sun. Due to land availability and the inland heat, we find that only 10% of the world's population exists in this region that encompasses 50% of the Earth.

The Earth works in cycles, some are daily, others seasonal, while others affect the planet every decade and beyond. The most obvious cycle is the planet's daily revolution that alters the day into night allowing no less than 12 hours of cooling every day when averaged over the course of a year. In a spinning sphere, the area at the top and bottom do not spin but rotate and the speed slowly increases as you travel toward the center. We use the 24-hour clock to measure this, and with a diameter of more than 24,000 miles at the equator, we can determine that the speed is over 1,000 miles per hour. This is the primary cycle that the planet uses to expel additional heat absorbed during the day and allows cooling during the evening. From a thermodynamics perspective, it continually refuels our planet into a state of chaos that is necessary to generate our daily weather systems as the cool night air meets the sunlight in the morning. Every time the planet turns into the sun, the heat rises and this rising temperature upon the surface increases weather activity.

The Earth's cycle, involving the axial tilt that alters the seasons, is a vital component of our planet's heating and cooling system. This action allows for a balanced distribution of yearly heating and cooling along with slow methodical adjustments.

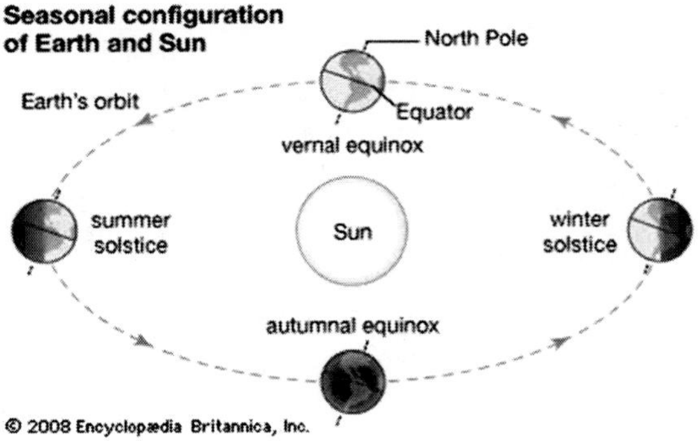

The planet orbits around the sun in an elliptical orbit and tilts on its axis, creating the alteration of our seasons. When the earth is closer to the sun, this is called the winter solstice, and the earth is tilted so that the southern hemisphere experiences summer while the northern hemisphere experiences winter. When we are farthest away from the sun, this is called summer solstice, and the planet's tilt reverses the seasons and the southern hemisphere experiences winter while the northern hemisphere goes into summer, even though we are farther away from the sun.

The tilting axis is vital for the cycles, seasons, and life on our planet. The Earth's axis tilts 23.5 degrees during the course of a year. To give a better understanding, this comes to a speed of 0.37 miles per hour, or about four city blocks per hour. If you were to start out on July 21 and begin walking south at 8.79 miles per day from Sioux City, Iowa, you would arrive in Mexico City on December 21. Along your journey, the sun would rise and set at the exact same location on the horizon, and your daylight hours would remain unchanged, even though you end up over 1,600 miles away.

This slow methodical tilt allows for the heating and cooling to be distributed evenly between both hemispheres every year, slowly adjusting for changes. The hemisphere begins to warm when there are more daylight hours than night, which averages March 21 in the northern hemisphere and September 23 in the south.

This tilting of the Earth's axis continually refuels the deep cool oceans with cool freshwater-generating circulation. Saltwater is heavier than freshwater, yet the water in the deep oceans have less salt content than the surface water. The primary reason behind this is the circulation and temperature varies as the cold freshwater stimulated every year from the spring thaw melts the snow and ice refueling the oceans with cold freshwater. Additionally, on the surface, we find water evaporating, leaving behind the salt creating a higher saline content. As a result, the cold water has a greater effect on the water's density than the salt content of the water.

There has been much speculation and even a recent movie titled *2012*, where the theory of earth crust displacement was presented. The magnetic pole is moving north, but has been moving since the early 1900s. The moving pole does not mean that the magnetic field is altering the land surface or the planet's rotation. There is one way to determine changes in our axial shift based upon known mathematics of a sphere. The Earth is round so if the axial tilt were to change, our sunrise and sunset times would alter too. Because our weather alterations occurred in the 1930s and today, and both sunrise and sunsets have been stable, then this theory is removed from having any possible impact on our current weather changes today.

The magnetic fields reach deep into space and are at their farthest point at midlatitude. This is also where the sun's radiance is the most intense for the greatest period of time. The magnetic field is vital to our planet, creating a shield that intercepts harmful incoming solar radiation from reaching the surface and creating an invisible shield that protects our planet. This field is visible in the far north and south where they generate both the southern and northern lights. Although this field may be altering slightly, it is constant and does not correspond to our oscillating weather patterns and timelines of the movement of the magnetic field. This eliminates this effect as a cause for our current climate changes.

The solar radiance's impact upon a sphere is going to be the greatest along the mid-latitude due to the length of time exposed to the incoming solar rays. This will generate the greatest amount of thermal radiation, causing an area of constant heat. This heat causes the atmosphere to swell like a ring around the planet. Although we see the seasons coming and going, what is actually happening is the planet is shifting into this heated ring and then moving out. This tilting action is what generates our seasonal changes. It is not our distance from the sun that causes these major alterations, although it does have an effect, it is the amount of time the area is exposed to the light that creates these changes.

Oceans

Around Antarctica, the oceans are able to flow continuously unabated by any land. It connects the Pacific, Indian, and Atlantic oceans in one circular movement from west to east. This creates some of the harshest ocean environments on the planet, but also maintains a cooler environment while the planet orbits close to the sun during their summer months.

The amount of water available in the oceans of the southern hemisphere allows the planet to consume and release much of this energy between seasons, creating a less extreme environment when compared to areas in the north along the same latitudes. A comparable understanding would be the difference between Washington DC and Melbourne, Australia. Both are coastal cities near the same latitude on the eastern edge of their continents, yet Washington DC averages a high of 89 degrees in the summer and lows averaging 29 degrees in the winter. In comparison, Melbourne averages a high of 78 degrees during their summer and an average low of 43 degrees during their winter months. Although the land is distributed unequally, the areas that do provide adequate rainfall and where irrigation can be used, double crops are the norm. This increases their production per acre compared to many areas in the north.

Fires and Dust Storms

As heat intensifies, the soil becomes dry and arid; fires frequently break out, expelling a great amount of ash into the atmosphere and begin to blow in the winds. As the water vapor rises and comes into contact with debris and dust in the atmosphere, the water vapor attaches itself onto these particles when enough gather together a cloud is developed. When saturated, it expels the water onto the surface, offering both cooling and moisture to the soil and vegetation.

Fire is a natural part of the planet's cycles, and in many areas of the world, the vegetation has adapted for this. Lightning, volcanic activity, falling rocks generating sparks, or spontaneous combustion in the right environment can cause natural fires. Lightning strikes the planet up to three billion times per year but is generally accompanied by storms. The storms provide quenching rains that will frequently extinguish the fires shortly after igniting. In some areas, like southern California and the prairies, this was part of its cycles, and we find the vegetation has oftentimes adapted for this. The rapid fires in the prairies assured the prairies remained treeless and would sweep across the land burning up all vegetation. The prairies' vegetation adapted by rooting itself deep into the soil, up to 15 feet. During dry years, the grass provided fuel for rapid moving fires, but the fires could not sustain for any period of time, and the fuel would be exhausted quickly. This would maintain the natural vegetation's root system, allowing for a rapid growth of new surface vegetation from the original roots of the plant. Saplings and other vegetation that would attempt to root into this ecosystem would die off, resulting in a treeless prairie.

Fires can be the preamble to dust storms that can cause great cooling. The fires in Russia in 2010 consumed vast amounts of land and vegetation, and resulted in the loss of many lives. Although the fires can cause short-term cooling through the smoke they give off by blocking the sun's rays, the carbon dioxide emitted and the loss in vegetation will have a more severe warming effect later. Man today is responsible for at least 90% of all the wildfires around the world and probably more. Man has wiped out vast amounts of land due to his activity on the planet, many of them considered accidental. This depletes even more of the Earth's resistance to solar radiation for many years to come, altering vast portions of the planet.

When dust storms occur, it's a sign of the planet's heating to levels beyond its natural ability to endure and can be measured by the loss of topsoil. When the natural resources do not alter and the land remains cleared for years, the planet will place sand over the soil to blanket and insulate the sublayers. This is what the planet was doing in the 1930s and what it is trying to do today. This sand heats and cools quickly and allows for a greater deviation of temperatures as a result providing cover for the barren land.

Atmosphere

We need to understand that global warming is a vital part of our lives every day. When the planet rotates and the day turns to night, if the warmth of the day was not retained, the planet would drop to subzero temperatures. This action is a result of the atmosphere and its ability to retain warmth. This is a vital mechanism for life as we know it on our planet today. It's for this reason that the term global warming often becomes controversial. What is being experienced is an alteration in the planet's climate trends and weather cycles where the overall effect is warming.

For simplicity's sake, there are only three reasons why we are not freezing in the same temperatures we are on the moon—that is, the combination of the sun, core, and our atmosphere. Altering any of these can and will have an effect on our planet's temperature.

As the Earth rotates and encounters the incoming solar radiance, meeting the cool morning surface, atmospheric disturbances develop. This is why at night and early morning, the lakes are frequently silent and often appear to be a sheet of glass but as the sun rises, soon the winds pick up and the still waters gradually shift to waves. Understanding this, we can also ascertain that the hotter the sun and earth, the more intensified the winds and storm activity will be. We can summarize that knowing how hot the planet has become and that solar maximums do affect our weather and heat, that both the amount of storms and their intensities, along with precipitation would naturally rise.

Increasing temperature is a delayed response to the amount of heat and energy retained, a by-product of the changes that have already taken place. In order for this to change, the actual temperature of the rock, soil, and water must increase or decrease which takes time and multiple seasonal cycles to alter. When the land warms, it will retain it longer, like having a fire and then adding rocks to it so the warmth resides long after the fire is out. This is what we are experiencing around the world—a longer delay before entering into winter along with an earlier spring thaw with increased summer heat.

If our planet did react immediately, then the warmest days would be on the days the sun is out the longest, and in the northern hemisphere, this occurs on June 21. Yet, we know that July 21 tends to be a much warmer period of time even though we are at the farthest point from the sun on July 3. Additionally, the coldest days are not on the shortest day, December 21, but later in the month of January when we are actually closer to the sun. This can also be seen on a daily basis: the warmest part of the day is not when the sun is at its peak, but afterwards. There is always a delay in temperature rise or fall after an effect takes place. The amount of time and intensity of the sun has a greater effect on our climate and temperatures than the Earth's distance from the sun, although they both play a role. Given a long enough period of cooling or warming in either the northern or southern hemispheres, there will be a gradual impact on the opposite region through heating and/or cooling transfer through the ocean's jet streams.

Our weather is greatly impacted by the fine balance between the Earth's orbit, the sun's intensity, the atmospheric content, and the location in which we live on the planet. The sun still rises and sets within the same time period as it has in the past, so the planet shifting can be ruled out as a variable in climate change. Although the North Pole has been moving toward Siberia at the rate of about 35 miles annually, if the Earth did shift, we would see a change in the sunrises and sunsets; in Minneapolis, this has not changed since the beginning of records. Although the magnetic field of the Earth is moving and a decline in the magnetic field can cause additional warming, this would not account for the cooling that occurred during the 1960s and 1970s, then returning back to warming once again. If this were the root cause of our altered weather patterns, then it would have remained constant throughout the last century as it began shifting in the early 1900s. Additionally, it would be recorded in Russia that their winters would be getting longer or more severe if there was an impact, but the opposite is true. Siberia today is experiencing the same delay going into winter that we are in North America. This fact rules out the magnetic poles shifting as a root cause.

After solar radiance passes through the magnetic field of our planet, the next encountered layer of this energy will be the ozone layer. Although science indicates that the ozone layer has been depleted in the past, science today shows that this layer has been improving and the alteration in our climate does not correspond to the time frames of depletion. Depletion would result in less resistance to solar radiation. This would increase the solar radiance upon the ground surface, resulting in a rise in temperatures over time. Because this occurred during our years of air pollution when the temperatures declined, this would eliminate the ozone as a cause behind our warming today. Since this time, science indicates that the ozone layer is healing from our early abuse of chemicals, returning to its natural ability to block incoming harmful radiation. This activity does not correspond to the oscillation we have seen in the temperature variables indicating that this layer is now maintaining stability.

The same is also true for carbon dioxide. Science indicates that carbon dioxide and the rate of warming was rising well before the heavy use of fossil fuels, as early as 1850. Carbon dioxide was debated as one of the primary causes for warming prior to the great Dust Bowl of the 1930s, yet was minimal in comparison to the levels of carbon dioxide we are releasing today. Then temperatures declined in the 1940s through the 1970s while carbon dioxide levels rose drastically. This indicates that carbon dioxide was not the cause and can only be seen as a by-product of another factor. The use of fossil fuels is rapidly accelerating our problems worldwide, but cannot be classified as the source of our problems today. So the question becomes, could we have altered the planet in such a way that the planet is now responding differently to the incoming solar radiance?

The atmosphere, its interaction, and the unique characteristics of water together act as a form of a natural radiator, water pump, purifier, and humidifier for our planet. As heat rises, so does evaporation and particulates, resulting in a rise in cloud cover. The cloud cover and rains offer a far greater impact on our temperatures and weather than carbon dioxide levels.

All of our heat comes from the surface of the planet as solar radiance is converted to thermal radiation—the heat we feel on the black top on a sunny hot day in our bare feet. Once this heat leaves the surface, it won't come back. Instead, it will build in the atmosphere and then can be carried by the winds to other locations. This is the most frequent way we feel warmth and change of seasons—through the warm and cold fronts that develop and move through.

Currently, the northern hemisphere is experiencing faster warming trends than Antarctica in the south. Some portions of Antarctica are actually increasing in size. The area of declining ice in Antarctica and the water that feeds the Arctic is influenced by the Atlantic Ocean. We know there are two variables that exist between the two hemispheres, population and land distribution. Heat in the north would impact the smaller of these two oceans, the Atlantic, in comparison to the much larger Pacific. The Atlantic currents impact the region of Antarctica that is melting and explains this differentiation. The Atlantic is also the primary feed for the Arctic Ocean. The Atlantic also has the Mid-Atlantic Ridge, a mountain ridge with highly active volcanoes where a gap between the plates exists.

The next layer the solar radiation will reach is the lower portion of the atmosphere, where we reside. Because all possible influences up to this point have been eliminated as the potential cause for our alterations, then this is the area in need of further exploring. This includes carbon dioxide. There was still one daunting question on my mind: if carbon dioxide was rising and temperatures were falling in the 1960s and 1970s, is there a source that has a greater impact on our weather than carbon dioxide levels, and if so, how could I prove it?

The Moon

Looking down at our planet from the North Pole, our planet spins and rotates around the sun in a counterclockwise direction. The moon rotates around our planet every 27.32 days and also rotates around the planet in a counterclockwise direction. The gravitational pull between the Earth's rotation, the sun, and the moon draw the upper level winds across the planet from west to east.

The moon comes up in the east in the early evening and by morning will be in the western sky. This gives an appearance as if it is moving from east to west, but is actually moving west to east but at a slower pace than our rotation. This creates the lunar cycles and the gravitational pull upon the Earth, affecting the tides, orbit, motion, and jet streams of our planet. The moon's orbit could indeed change our weather if altered, but has remained constant with no significant alterations that could cause climate alterations that Earth has been experiencing.

The Sun

The sun goes through cycles averaging approximately every 11 years. The scientific community has at times minimized its impact on our weather, but as records will clearly indicate, this is the planet's primary influence upon its cycles and surface weather. There is not enough energy within the core to heat our planet without the sun. Without the sun, our planet would freeze and die. The question then becomes, how is our planet responding to the sun? We know that man cannot change the sun's radiance, nor has the radiance of the sun changed dramatically. It has remained constant over the years, running on a series of cycles.

These cycles, when interrupted, will produce predictable effects upon our planet. These cycles, such as the Dalton Minimum from 1790 to 1830 and the Maunder Minimum from 1645 to 1715, are periods of time when the sun's normal activity declined for unknown reasons. Named after the scientists credited for noticing these changes, the Dalton Minimum can be seen during the time frames of 1790–1830 in the following graph.

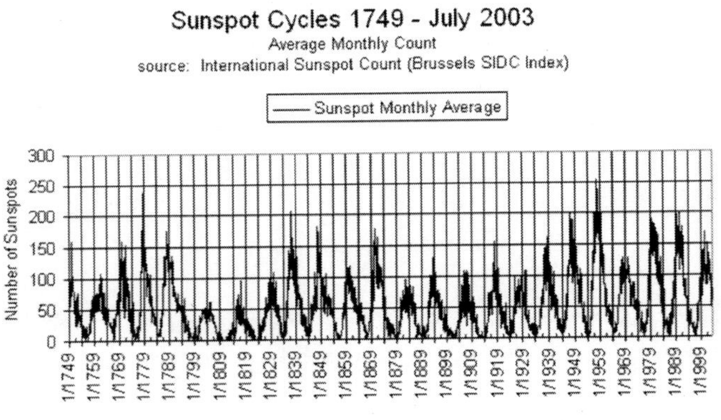

The solar cycles do have an effect upon our weather, but due to increased solar output during the 1950s and 1960s in comparison to the 1930s, it indicates that the sun is not responsible for our current alterations and remains on a semipredictable and constant cycle.

The Core

The core of the Earth is a sealed system, and there are many unique characteristics associated with this. Heating within a sealed system has different properties than heating in an unsealed system. When we heat a sealed container, our first concern must be the strength of the container to hold the pressure that the heat creates. As a result, thermodynamics applies unique laws to this form of heating.

Granite begins to melt at about 2,219 degrees Fahrenheit depending upon pressure and composition. In general, science has determined that the temperature rises 1 degree Fahrenheit per 70 feet in depth once you pass through the stable temperate layer on the surface away from areas where the plates have contact. Volcanoes and hot springs confirm the heat coming upwards and deep mineshafts help support this heat gradient. Based upon these figures, we can determine that granite would begin to melt at a depth of approximately 29 miles beneath the surface. Due to the increased pressure at this depth, the melting point will naturally rise, driving the melting point even deeper. To give us a visual perspective of this layer, if you were to drive from downtown LA to the Los Angeles International Airport, you would have traveled the thickness of our crust. If you then boarded a jet and traveled to Jerusalem, you would have traveled the distance to reach the other side of the Earth. As you can see, the surface layer, or crust, is a very minimal section of our planet in comparison to the whole.

Early science indicates that the core, left over from the formation of the planet, would eventually cool and was seen as insignificant. Although science continues to discount the core's ability to offer significant heat for the planet, it could cause additional heating if altered. The core has been maintaining a stable environment and cannot account for these oscillations in our weather.

These two sources, the core and the sun, are the planet's only sources of heat. If the sun and the core have remained stable in their cycles, then these heat sources cannot be a contributor to our current climate alterations. If the heating has remained constant, then we need to look to alterations that would cause a change in the way our planet absorbs and releases this heat. Carbon dioxide is one of these elements and is now easy to see why the science community has remained fixated on this issue.

SEARCHING FOR THE SOURCE

After researching for several months into the possible causes for these changes, I decided to take a walk in the forest. It was now the spring of 2010, six months into my research, and the solitude of the forests in the spring after the winter months is so appealing.

As I looked up into the sky, my thoughts began to revert back to the earlier days—the 1960s and 1970s. This was the time of the Vietnam War, but also a time when there were marches on Washington demanding clean air and a reduction in pollution. The air back then, especially in the cities with all the trucks and buses burning diesel fuels would often burn my eyes. I was recalling my mother's earlier conversation of those early days of smog and the use of coal. This changed after World War II, and we went to natural gas for heating; but our driving habits and industrial growth resulted in a continual increase in air pollutants. The conversion cleaned the air, but we began to replace it with other contaminants, especially from the automobile.

If pollution caused a restriction to incoming solar energy and this resulted in cooling, then obscuring the sun's radiance upon the surface could be one of the means of reversing our current heating. But without finding the source, this would mean living with obscured sunshine forever, just like it was when I was young. Life as it is in Beijing, China, for our children. Reverse the EPA's clean air standards or find an element that would not result in the environmental troubles that early air pollution caused, and it would create a cooling effect upon the planet.

This would also mean that the polar ice caps would have grown to levels during this period beyond its norm, meaning a receding polar ice cap would be natural as man's activities had actually increased its size during these years. Then the question would be what would be considered a norm for the polar ice caps? We know that the Northwest Passage was first navigated by boat by the Norwegian explorer, Roald Amundsen, but it took three years to accomplish, from 1903 to 1906. As a result, the Panama Canal was created. By understanding this, we know that the passage opening as frequently as it has over the past decade indicates we have warmed significantly today.

As I continued walking through the forests, my mind began to process our planet's heat. I walked out of the forest and the sun began to beat down upon my face, and in the springtime, this feels so warm and inviting after months of cold and being cramped up inside. It was very clear that the sun's impact with an object, such as my face, increased heat. I then lit up a cigarette, something that I have recently given up, and began to stare at the flame. All the heat was rising and nothing comes down, meaning that all the energy rises to a cooler environment with lower pressure—this is a basic rule of thermodynamics. It can rise up and build and then move where the winds will bring it, but it is released from the surface immediately. As I looked down at my feet, I realized that all of our heat is coming up from the thermal radiation produced by the sun's light coming in contact with an object. I thought of the core, and the same applies for this. All the heat from beneath us is slowly trickling up to the surface, because heat rises to cooler areas that are under less pressure. I began to quickly understand that the sun is like our boiler, the primary source of heat for our planet but doesn't directly heat it. This energy is then delivered through space and our atmosphere, just like water lines delivering energy to all the rooms in a home. When this energy impacts an object, then that object becomes the radiator, the source of our heat. This means that all of our heat comes from the ground and is impacted by the solar radiation, and then the object emits thermal radiation, increasing our temperatures. If a particle of solar energy passes through our atmosphere without coming into contact with any other particles, it will continue to pass through and have no direct impact on the warmth and temperatures that we feel at ground level. After all, the temperature levels and weather as we know it is between the ground and the first several feet above it. We may be at different elevations, but within the same proximity to the ground.

I decided to return to my truck and get a temperature probe that I use for work and began taking temperature samples. It was a sunny day, windy, and 79 degrees. It was just afternoon and I found the warmest was the blacktop at 105 degrees and was transmitting the heat outwards into the surrounding soils several feet away from the surface. It was acting like a large heat sink, absorbing this energy. The open farm fields were 100 degrees with the heat penetrating several inches down into the soil. The prairies were 76 degrees and reduced to 70 degrees within three inches of the surface.

I then went into the forests and everything became very clear. I found that the soil surface that is hidden from view and resides below the thatch never exceeded 71 degrees and was a stable 69 degrees several inches down. Thatch is the leaves and debris from the previous years accumulated on the ground found in nature. This thatch was maintaining a very cool damp environment throughout the entire forest floor even where direct sunlight was available. The thatch insulated the ground from direct contact with the sun. Nowhere in the forests or the prairies is there direct contact between the sunlight and the soil surface. In these areas, only a washout or a forest fire could expose the soil to direct sunlight, but even in these areas, the period of time is very short-lived as new life spawns rapidly. Within these biomes, there is no sunlight that reaches the soil surface. This was something I had never thought of, an insulating cover over the forest floor. This explains why that pile of leaves we never got off the ground before it snowed was still filled with ice when we went to rake them up in the spring. Good insulators slow the transfer of energy, and the multiple levels of leaves with air between them are an excellent insulator.

I returned with the results and took down the official temperatures. Thermal equilibrium is the ultimate goal of heat, to become one stable temperature. If humans were not present, it is clear that the air temperature on this day would have been between the temperatures of the prairies at 76 degrees and the forests at 71 degrees. This would place the temperature on this day about 74 degrees. Because the forests and prairies are now a small percentage of the land and more than half has been converted to farmland and urban development with temperatures of 100 degrees or more, this variable has to be applied to the equation. This effect was clearly the cause this day, driving the air temperature up to 79 degrees, several degrees above the warmest natural environment, the prairies.

The sun itself can only penetrate so deep into the Earth's crust, but it is clear that rising temperatures during the spring is one source of increased heat, one cycle that mankind has altered. When the crops mature, they begin to shade the surface with their growth. This aids in decreasing this variable in temperatures, but without the thatch, there would be some differentiation, especially with the surface moisture. How it all works was going to take more time to understand, more temperature samples to take and more research was necessary.

It's human nature to look to the extremes in temperatures. When it's summer, we look to the peak heat of the day, and in the winter, we look to the extreme low temperatures and hope for the highest. When looking to global heating and cooling, we need to look at the ground temperature, the source of our heat, and this means looking at the low temperatures at night.

When the ground becomes free of frost, the air temperature becomes controlled by the ground. This becomes evident when the low temperatures begin exceeding 32 degrees and the lakes become ice-free. At this point, the low air temperature at night is a reflection of the heat applied to the surface as well as the latent heat below. The daily high temperatures are ambient. They come and go, but ground heat remains for a period of time and cools once a year. Because of the change of seasons, it requires the passing of many cycles (years) to alter.

Another principle governing thermal properties is that when given two heat variables in a neutral environment, meaning both will affect the surrounding area temperatures, the temperatures will be higher than the lowest temperature, but lower than the highest.

Isaac Newton's law of cooling explains why our coffee gets cool and why our milk gets warm when they sit out for a period of time. This also explains why we get home from work, and the temperature in our home may be hot and it takes time to cool, or if it is cool it takes time to warm. The time variable is determined by the size of your heating or cooling source, the area you are altering, and the insulating value of the object from the exterior environment. All of these are factors in heating and cooling. No matter what your temperatures are, thermal equilibrium is the final outcome of heating and cooling and given enough time will ultimately come to one common temperature.

On June 6, 2011, the recorded temperature in Minneapolis was a high of 97 degrees. I had determined that the greatest alterations were in the spring after the ground surface had been exposed to the sun's rays for a period of time. The temperature this early in the year would be ideal for displaying the temperature gradients and their effects.

The air temperature at the site was 94 degrees at 1:00 p.m. The blacktop roads were 133 degrees. In contrast, the farm fields were at 124 degrees. I then took samples of the soil in the forest and found they averaged 74 degrees while the prairies were 78 degrees. According to Isaac Newton's law of cooling, we can determine that the temperature will be less than 124 degrees (calculating over 50% land alterations) and greater than 74 degrees, depending upon the size differentiations and the time allowed to heat. This accounts for the temperature being 94 degrees this day. The following day, it was 103 degrees. Before man had come to this land in the 1800s, all the land would have been forests and prairies, meaning the air temperature variables would have been more than 74 degrees but less than 78 degrees, and it would not have been possible to be 94 degrees on this day.

During the summer months when the ground is thawed, the temperatures of the early morning match the temperature of the ground. As the sun beats down upon the Earth day by day, the temperatures gradually rise, and when they become overcast, the temperatures gradually fall. Because our heat comes from the ground beneath us, it is important to understand how this may have been altered.

On July 2, 2012, I placed a video online explaining these principles and findings.[1]

US Department of Energy:

> Solar heat absorbed through windows and roofs can increase cooling costs, and incorporating shade from landscaping elements can help reduce this solar heat gain. Shading and evapotranspiration (the process by which a plant actively moves and releases water vapor) from trees can reduce surrounding air temperatures as much as 9° F (5°C). Because cool air settles near the ground, air temperatures directly under trees can be as much as 25°F (14°C) cooler than air temperatures above nearby blacktop.[2]

EPA:

> Trees and vegetation lower surface and air temperatures by providing shade and through evapotranspiration. Shaded surfaces, for example, may be 20–45°F (11–25°C) cooler than the peak temperatures of unshaded materials. Evapotranspiration, alone or in combination with shading, can help reduce peak summer temperatures by 2–9°F (1–5°C).[3]

In these researched statements that I have confirmed as true, scientists indicate that the increased solar radiation placed upon our homes will increase the thermal absorption, resulting in more heat and driving the temperatures up. This is the connection that I believe is evading science; our houses are built on this planet, and this planet is our home. If this affects our houses as they say, then it will also affect our home, the Earth, resulting in rising temperatures.

If we understand this philosophy, it is easy to understand that solar heat is absorbed at a greater rate when the forest, prairie, and thatch cover is removed. This in turn will emit more thermal radiation, increasing the heat and raising the surface temperatures. The shade they provide reduces the solar heat gain and provides a significantly cooler air temperature. Additionally, they block the wind and allow for a cooler environment on the ground level.

When I researched North American history, I found that early settlers said the forests were so dense that a squirrel could jump from tree to tree from the Atlantic to the Mississippi and never touch the ground. It was also stated that a squirrel could jump from nutcracker tree to nutcracker tree from Massachusetts to Georgia and never touch the ground. This is important in understanding the amount of alterations that have occurred over the last few hundred years.

I then began to think about the carbon dioxide curve and the noticeable rise that began in the 1800s that has continued to this day. I looked around to the forests and everything was budding, but the farm fields were barren and desolate. As I walked through the fields, I also noticed they were free of insects. The process of photosynthesis only occurs during the day in warmer months, and it is this process that removes carbon dioxide from our atmosphere. This period of time peaks in the springtime, and this occurs from April to June in the northern hemisphere. The beginning of the rise in carbon dioxide levels seemed to correspond to the same time periods we desolated the landscape during the 1800s.

This thermal energy begins to rise and build during the spring when daylight hours exceed darkness, and declines in the fall as the daylight hours decline. Different elements in our atmosphere such as water vapor and carbon dioxide restrict its flow and help maintain some of this heat within its biome. The Earth itself at the point of impact is where the heat radiates from, and if I wanted to find the storage site for the rising heat, it was clear that the low temperatures at night are what I needed to focus on. This gives us an indication of how much energy is held overnight (heat storage) before the next day. If it's a sunny day on July 4 in the northern hemisphere and the morning begins at 72 degrees, it will be a much warmer day if the day starts out at 80 degrees.

Then my thoughts turned to the Dust Bowl and all the sand during this period of time. I began to think about time I had spent in the deserts, and I realized that during the heat of the day, the temperatures would rise greatly, but nightly they would drop drastically. This indicates that the sand was a material the planet used to cover itself when the surface is exposed to intense solar radiation without any shade available. Like a blanket, the planet covers the exposed soil and rocks with sand to insulate it from the incoming daytime heat, then exhausting this during the evening. This explained the Dust Bowl era as the planet was covering itself from the incoming solar energy with a blanket of sand. It's one of many automatic responses from the planet to cool itself.

When I returned, I began to explore the era of the 1920s and 1930s when the scientists were making claims of carbon dioxide being a warming agent, but were silenced by the cooling that occurred after World War II. This is further justified by the amount of cooling that occurred in the northern hemisphere where the majority of land, population, and industry thrived in comparison to the southern hemisphere. This would naturally require several years to build in the atmosphere that must have occurred coincidentally during the 1930s. It took time to build in our atmosphere, and it would also require time to implement and improve our air quality. This explains the number of years necessary to alter our planet's temperatures by heating, cooling, and then heating again, and accounts for the oscillation.

This observation does strengthen the carbon dioxide argument with two exceptions. First, the air pollution and the resistance to incoming energy had a greater impact on our planet's temperature than carbon dioxide levels in our atmosphere. This clearly indicates that the amount of solar radiance reaching the surface has a greater effect on our surface temperatures than carbon dioxide levels. And second, carbon dioxide levels from emissions can't explain the warming that was building from the late 1800s until it climaxed in the 1930s during the great Dust Bowl. Carbon dioxide levels were on the rise 50 years before large-scale use of fossil fuels even began. This alone pointed to a root cause that has yet to be explored. Carbon dioxide graphs universally demonstrate a continual rise in levels prior to the 1900s.

High carbon dioxide levels do increase the insulating effect of our atmosphere and impacts heat flow, but any form of rise would be slow, methodical, and evenly felt based upon my experience and knowledge of warming. The rate of rise we are experiencing doesn't correspond to my knowledge of how adding additional insulation over our planet would respond, nor does it coincide with what we would expect within the environment of a green house. Temperatures should increase at a more rapid rate as we rise in elevations due to the rise in restriction to outgoing heat and additional carbon dioxide levels impacting this as we rise in altitude. Like adding insulation in a multistory home, the warmth will build and be felt as we travel up the stairs.

If man altered the surface by removing the planet's resistance to the incoming solar radiation, we would heat up and carbon dioxide levels would rise. This would account for the previous heat of the 1930's great Dust Bowl. We polluted the air and increased this resistance, which accidentally offset the loss of surface resistance and we cooled, even though the carbon dioxide levels were still rising. Then we enacted the EPA and cleaned the air, and we reheated again, and we are beginning to see the return of the Dust Bowl era of the 1930s.

While researching, there are two events in particular that began generating public awareness of the problems with air pollution. There is a small town called Donora, Pennsylvania, where the impacts of air pollution killed forty people in December of 1948 and left half the town with respiratory problems. Later, in December of 1952, London experienced one of the worst peacetime disasters when a killer fog engulfed the city killing thousands of people.

Due to massive protests sweeping the nation, the country was compelled to act and the Clean Air Act was instituted in 1967. Later, the EPA was established and sweeping changes began in 1970. Most notably in the early days was the switch from leaded to unleaded fuels and mandatory pollution controls on the auto industry that began in 1972. One by one, industry by industry, the EPA gradually cleaned the air, allowing more solar radiation to reach the surface. Not only did they clean the air, but the lakes and waterways along with the land. It appeared to me that there was a buildup of atmospheric pollution, offering enough resistance to the incoming solar radiation to offset the warming experienced in the 1930's Dust Bowl era. If this is true, then we need to look back to these years and beyond.

I decided to do more research into many of these questions and began to run statistics on our low temperatures that would determine the amount of heat storage present in the earth year by year. In the graph "Minneapolis July Low Temperatures 1891–Present," there is a dip in temperatures and seasonal alterations indicating a cooling period from 1940 until the 1970s, identical to previous data curves. Today, science has confirmed that the amount of pollution in the atmosphere that man had inadvertently built up over the years had actually initiated a cooling trend. In 1970, the

EPA was developed. In 1972, we switched to unleaded fuels, and after years of changes, the skies have become clear once again. All of these graphs indicate that 1996 was the year that our warming began, but we must also remember that due to the Earth's cycles, both time and temperature are relative. This is not a pinpoint time of its beginning, but the beginning signs of warming. Like turning your furnace on when it is cold, after a period of time, you begin to notice a change, but this takes time for thermal equilibrium to take effect and will eventually be noticed by a rise in temperature.

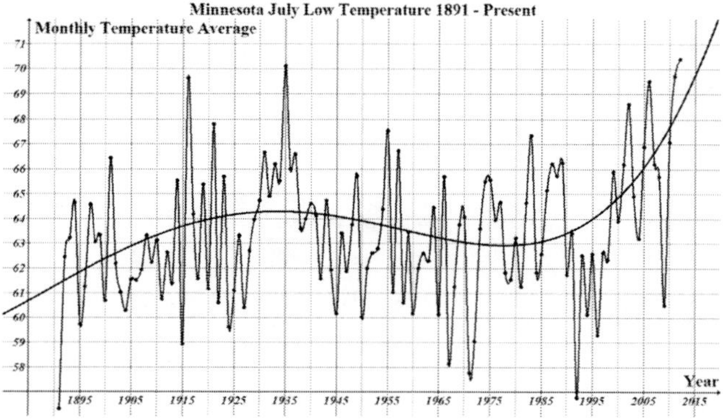

By looking at these temperatures we find that 2012 was 70.4 degrees, over taking the 1936 record of 70.12 degrees in Minnesota. From 1891 until 1899, the average temperature was 62.06 degrees, 8.34 degrees cooler than 2012.

I decided to look to a higher elevation and determine if these rising temperatures were different. For example, if we increase the insulation in an attic, the upper level rooms will increase in heat at a greater rate due to the increased resistance to outgoing heat and will affect this environment greater than the lower levels and will be registered by the temperature differentiations. If temperature variables were notably greater at a higher elevation than lower elevations, this would give carbon dioxide credence, proving a greater variable restricting outgoing heat. When compared to the statistics for this era in Fort Collins Colorado,[4] we also found the same results. From 1889 to 1898, the average temperature was 52.82 degrees, and in 2012, they experienced an average low temperature of 61.7 degrees, 8.88 degrees cooler than 2012 and the warmest on record since the year 1889.

The warmest year in Minneapolis prior to 1900 was in 1894 when an average low temperature at night was 64.64 degrees, 5.82 degrees cooler than 2012. In Fort Collins, it was 1896 when the low temperature was 55.6 degrees, 6.1 degrees cooler than 2012. This confirms that the warming is across the continental plate and the same at different altitudes.

Because the planet was already warming by the 1890s and deforestation had begun in the 1830s in Minnesota, it is estimated that the average low temperatures at night in Minnesota would have averaged as low as 60 degrees at night before man drastically altered the landscape. This would place the temperatures in Minnesota in 2012 over 10 degrees above its natural state today.

Since I was young, I used to wonder what the world was like around me before America was colonized. I have heard many people state that we have more trees today than what we used to have, but I always question the starting date they are using. I look back to the 1803, the land Lewis and Clark would have seen. After the Revolutionary War and the Louisiana Purchase in 1803, logging went unrestricted for the next 120 years before any action was taken. Because we are returning back to the great Dust Bowl, I decided to begin to look back in time and try to understand what transpired. Whatever caused this event is the same catalyst for our current alterations based upon the understanding of global dimming. I decided to do some historical research and determine what alterations man may have made to the land that could have caused this event.

Historic Timelines for North America

After the plague rampaged the world in the mid-1300s, it took several generations to rebuild. With the finding of the New World, there were many explorers and missionaries who began to come to America. Early explorers brought with them diseases such as smallpox and measles that had not been introduced into the Native American population that decimated the population across North America in the early 1500s. This had a similar effect as the plague had in Europe and Asia. For the next 300 years, this land was allowed to return back to its original natural form. By the 1600s, 250 years after the plague, Europe's population was expanding beyond her resources and America became Europe's overflow. Many traveled to the new world offering prosperity, freedom, and hope.

After the Revolutionary War, England set her eyes on Argentina and Australia to support the needs of her empire. By 1800, every corner of the world that had reasonable climates were now inhabited, even many that weren't.

As I looked back into the history of the lower 48 states, there are many events that altered our landscape. In 1800, the United States was primarily occupying the eastern seaboard and the western United States was on the other side of the Appalachian Mountains, separated by a very dense forest with no roads connecting them. Fur traders and explorers were primarily using the waterways for transportation and trade. In 1803, the Louisiana Purchase opened up a vast amount of territory, allowing for further expansion. According to the Lewis and Clark's logs, Elk was a part of their diet along the Missouri River as far south as the Nebraska and Kansas border. They encountered there first herd of bison that they estimated at 500 in present day South Dakota and killed an estimated 227 along their journey to feed the team. Bison roamed the plains the same way we see the wildebeest in Africa today. The use of barbed wire in the 1870s stopped the migratory paths of the bison, and they were virtually wiped out before 1900 as the land was converted.

The land was parceled off and homesteading began redesigning the landscape into farmland. An overpopulating Europe, compounded with problems of their own, created a massive overflow. As a result, the government kept accurate account of the land use over the years, allowing us to see how this was shaped through homesteading.

By the mid-1800s, farms had spread west to the prairies, but advancement came to a stop. The prairies, with no lumber available, deep-rooted vegetation and damp soil slowed progress. The plows were getting bogged down in the soil and made the land very difficult to transform.

John Deere made his mark in history when he designed the rolling clipper plow that allowed farmers to till the soil without getting bogged down. The railroad soon came through, and we were now able to clear land rapidly with the industrial movement introducing items such as steam tractors, plows, thatchers, and eventually, the conventional machinery we see today. We could now clear land rapidly, especially in comparison to prior generations when this was done by hand. By 1860, the total amount of land use for farming was 21% of the entire lower 48 states.

The Homestead Act of 1862 was enacted by President Lincoln and was the first of many that began opening up large areas of land for farmers. Migration intensified as a result, and the people continued to flow into America as land was being offered to people for free that were qualified, willing to work the land, and lived on the land.

I had a great-grandfather and great-grandmother come over from Sweden and homestead a farm here in Minnesota in the 1880s. After the death of my grandmother, I inherited their wedding Bible and the wedding certificate in the Bible was dated March 23, 1889. It contained written family weddings, births, and deaths until 1987, 100 years of history. Curious about these early settlers who came and stripped the land, I decided I wanted to understand them more, and I decided to return to their homestead and donate their Bible to the church that was built by them and neighbors. I felt their Bible deserved to be with them and the church they had built, their final resting place. They are buried on the parcel of land donated to the church located in Dassel, Minnesota.

I met relatives on this day whom I had never met, and stories of the early 1900s. In the few pictures I have of them, I noticed she had a stern look upon her face. It looked as if she had led a difficult life. I asked my mother about her and she began to tell me a story that was told to her by her mother, my grandmother, that happened in the early 1900s. An influenza outbreak had occurred early in the 1900s and her mother's sister, along with her entire family, became ill. Many people were dying from this outbreak around the nation, but she decided to leave her family and go help her sister's family. My grandmother was only about nine years old when this occurred, and she told me that she recalled the fear in her mother's voice when she told her this story. She told me that the family thought they would never see her, my great-grandma, ever again. What happened at this point is probably the years of hardship I see in these pictures today. Her sister and husband lived, but she had to bury all nine of nieces and nephews. I can only imagine the pain in one's heart having to endure such a traumatic event. The father and mother were too weak to help her, and she would not allow anyone to come to the home and risk infection. For some reason, she never became ill.

As I entered into the original church they had built that was now being torn down, I could get a feel for what they endured. The church was small, the size of a dining room. Land, materials, and labor were supplied by the area residents that could find support, love, inspiration, and assistance during difficult times. It is in understanding this, the importance of family that my mother always spoke of, that I now truly understood. These early settlers came here to find a new life, freedom to choose their own destiny, a place of hope and prosperity in return for their hard work.

My grandfather had an older brother, and in the early days, the farms were inherited by the eldest son. He decided to leave Sweden and travel to America where he met up with another brother and settled here. He was one of multitudes that began to make a new home for themselves in the late

1800s and continued until reforms took place in the 1920s. Many of us are here today as a result of this migration.

Transportation of food was through the use of rivers and waterways. Due to the weight of rice in comparison to other crops, specifically corn, this became the predominant crop early in our history. Rice requires water that infiltrates into the ground and aids in cooling, whereas corn requires drier ground, exposing it directly to the solar radiance and also causes additional water runoff. Corn and other dry crops alter the land to move the water away from the fields. Without the vegetation in place and top soils cleared, water runoff became a problem. Soon, flooding began and levies were installed to channel the water. These levies were constantly being installed and risen over the years as we have seen in New Orleans, and today, we still need to raise them.

In 1902, Teddy Roosevelt instituted the Reclamation Act, damming up nearly every river west of the Mississippi for irrigation purposes. By 1905, in response to the loss of land and the potential threat this imposed, Teddy Roosevelt began sectioning off vast amounts of land to be preserved as national forests and parks across America. By 1914, 44% of the landscape was altered to farmland to support the war effort.

Logging continued vigorously across the United States until the 1930s when controls began to be implemented. Even today with the use of air-conditioning, clear cutting prior to construction is a natural process. Before air-conditioning was readily available, homes were built utilizing the shade of trees in order to provide a cooler environment during the summer months. When the leaves fall in autumn, they offered sun for additional warmth during the winter months.

By the 1950s, America reached its peak when 59% of the entire landmass was used for farm land alone. The amount of paved roads alone in America today could pave the state of Kentucky. And now add the urban areas into this, subtract the Rockies and deserts of the southwest, and we can see that nearly every inch of the lower 48 states has been drastically altered from its original state in 1600.

Research demonstrates that the temperatures are rising, and the source comes from the Earth beneath us. How it all works is going to require far more research, so I decided to look into volcanic eruptions for two reasons: eruptions are the Earth's discharge of massive amounts of carbon dioxide and looking into these would aid in understanding how the Earth naturally uses and balances carbon dioxide levels, and thermodynamics tells us that if there is a rise in heat, then there will also be a rise in pressure within a sealed system.

ERUPTION RESEARCH "PRESSURE"

Up to this point, I had determined that the heat was rising because we had literally cut the Earth down and keep it barren year after year. This causes a slow and methodical rise in temperatures and increased thermal absorption that increases year by year. All of this heat can be registered on the surface by a notable increase in low temperatures and their alterations over time. The loss of the forests and prairies were causing an early rise in temperatures in the spring as a result of the barren soil's exposure to the sun. This would naturally increase carbon dioxide levels and temperatures over time. The ground clutter, or thatch, removed the insulating cover and dried out the surface soils. Additional problems would come from water runoff, erosion, and degradation of the soil. The increased heat resulting from thermal absorption remained dormant in the soil longer, causing an extended delay in the seasonal shift to winter. With shorter and milder winters, earlier springs resulted. Alterations of the land surface explain much of the findings to this point and were causing the shifts, but how it was all linked together was still in question.

I decided to continue my search for how this heat is increasing in the ground beneath us. From a thermodynamic perspective, volcanoes are an ideal focus for research. Within a sealed container whenever heat rises, so does pressure.

As I looked at the volcanic data (VEI-4 and greater), I could see a pattern that develops around the solar cycles. Specifically during the lull in the solar activity that creates periods of known cooling, in particular the Dalton and Maunder Minimums, volcanic activity reduced greatly. These large-scale eruptions also increased around the period of time when the solar maximums took place, and had a tendency to increase either during or just after this event. Seeing these events rise during these cycles indicated a normal response from increased solar radiation when taken into account that heat, time, and temperature correlate with one another. Although I could see a pattern, it was like a blur of eruptions all mixed together. For months, I pondered these events and my instincts kept me looking at these eruptions.

One early morning, I was looking more in depth into these eruptions and decided to cook some eggs for breakfast and take a break. I had a bent up pan and the eggs had gathered on the edge of the pan that was not in contact with the burner and they were not cooking. As I brought the eggs back onto the burner, I had a thought.

If I had a large pan and only one burner was lit, the eggs would cook only under that burner. Then I imagined a large pan of water on my stove with only one burner on, the water above the burner that was on would begin to boil first. I cooked my eggs and went to work by separating the volcanic eruptions by their regions. Like reading a book, the results slowly revealed themselves.

Volcanoes: Earth's Historical Thermostat

Volcanoes are placed upon a scale according to their intensity and is called the Volcanic Eruption Index, or VEI. Only large-scale eruptions of VEI-4 and greater that could alter the planet's weather systems were researched and used as the baseline for this research. Any references to eruptions in this chapter are to be associated to only these large eruptions.

Eruptions of this magnitude alter weather patterns and cool the planet acting as a pressure relief valve. Specifically, when they alter the weather conditions on the surface, there should be a condition and cause for such an event.

Because temperatures were never recorded until Daniel Fahrenheit came up with a way to measure and record variables in temperatures in 1724, and become a standard over time, most temperatures recorded did not began until the mid-to late-1800s. Needing a more complete historical timeline and understanding the physics of heating a sealed container, this alternative method of research was selected.

The Earth has a molten core sealed within a surface that we call the crust. We know this because of volcanic activity, geysers, and hot springs. Because of the unique features that a liquid possesses when it is in a sealed container, the research was new and unique. To understand the differences, take two bottles of water and open only one of them and then place them both in the microwave on high and watch how they differ in their reaction, which most of us have done at one time. A liquid in a sealed container, even the Earth, increases in pressure until an eruption occurs. This is the principle of physics that we are capable of duplicating in the microwave. Pursuing this avenue of research revealed a vast array of discoveries that brought about many explanations to our planet's heating and cooling cycles and a past historical time frame that aids us in understanding these cycles and their causes.

Current research has aided greatly in the forensics of past eruptions, but records and accuracy become more obscured as we travel back in time. This happens as volcanic eruptions, time, and erosion all take their toll. Notable time periods would be prior to 1492 in North and South America, Iceland prior to 1874, Australia prior to 1788, and the northern Pacific region including Alaska and Russia prior to 1800. Many of these historical events in Asia and Europe have a written record, allowing scientists to examine these eruptions in greater detail. Information about the forensics involving the examination of past eruptions can be found at the Smithsonian Institute and is known as the Global Volcanism Program. Because this data was raw and unmanipulated from volcanologists, the accuracy for this review is an optimum database.

The data was plotted on timelines and separated by the tectonic plate associated with the eruption. Thermodynamics tells us that if you have a large container of fluid and heat only one area, the heat and pressure will be the greatest nearest the source of the heat. This indicates that if a plate was to heat for any reason, the pressure would be greatest under that plate and a volcanic eruption will occur at the weakest location to the source of the heat, typically along its plate boundaries. Like a pan on the stove that is sealed and heating, the weakest point, no matter where the location, will give way if the heat continues. By plotting these eruptions, there are clearly definable date lines where there was a distinct rise and fall in eruptions. These occurrences and their causes are what I examined.

This is a map of the Earth's continental plates and the directional forces being applied.[1] The Mid-Atlantic Ridge is located in the center of this map in the Atlantic Ocean, and this is where the planet is split and the energy is driven outwards from this location.

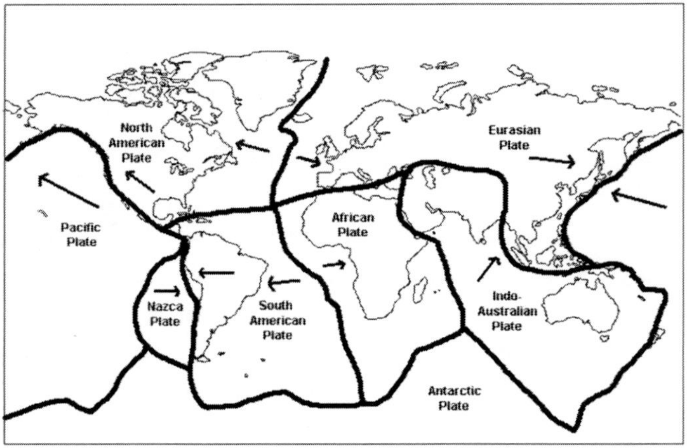

Volcanoes have always been a part of the planet's history. They offset additional heating that may result from an increased solar output, or perhaps a meteor wiping out a large portion of the Earth's land surface. They can be considered a pressure relief valve the same way you will find man using it to release pressure to ensure the safety of a machine. In regards to the Earth, they keep us from literally blowing apart.

As these records were plotted out, there were enough documented eruptions to begin, plotting the rise and fall in these large volcanic eruptions as far back as ad 700. Although there is further research that could be done going further back, there has not been any research as of this time, and further forensics and carbon dating would be needed.

All data is acquired from the Global Volcanism Program from the Smithsonian Institute and is the site that the USGS recommends.

In the following graphs, there are three periods of time that need to be considered. These periods of time were specific times of cooling due to the sun's decreased intensity upon the planet's surface, two natural and one man-made.

- Maunder Minimum (1645–1715)
- Dalton Minimum (1790–1803)
- Global Dimming from air pollution and its impact from post WWII to the 1980s

Eruption Rating

Eruptions within these magnitudes are rare, so they are divided into the total amount of eruptions every decade. All volcanoes are graphed per decade and having two eruptions—one in 1980 and another in 1989—will be displayed as two eruptions in 1980.

Calculating eruptions was based upon this decade long rate and then observing periods of time when increases and decreases occurred. An example would be if the North American Plate had ten eruptions within a 100-year period of time, this would result in a 100% chance of an eruption per decade. In the following 100 years, we find only five eruptions, making this an eruption rate of 50% per decade.

We can then conclude that the first 100 years were twice as active as the century that followed.

After the eruption data was plotted and periods of time became clear, I had one question, "What were our ancestors doing back then?" After all the times were plotted, I placed the following time frames, and as I focused in on these periods, I saw another pattern develop—a pattern within mankind.

North America Plate

When eruptions occur in this region, specifically near the Gulf of Mexico, they have a direct impact on the ocean's jet streams, affecting the weather patterns across the continent. They reflect a great amount of solar radiance back out into space and this in turn cools the Gulf Stream. Reducing this heat allows the Arctic jet streams to have more momentum, and this alters the inland precipitation and temperatures by altering the jet streams. This can also create major storms and flooding in some areas, while bringing droughts to others.

In North America, we have an opportunity to better understand mankind's responses to climate change by looking back to the 1930s. During this period of time, we did see hunger and massive migrations of people to other areas that offered necessary resources. This research and how mankind responded can help in understanding previous events along with the potential size of these civilizations and what would have happened.

In North and South America, the year that these large eruptions have occurred is known with 100% accuracy since 1800, and only one questionable eruption since 1500.

Before 1570, there are many uncertain dates averaging a plus and minus range of about 100 years. Because of specific groupings found and relying on the endless hours of scientists attempting to get these dates correct, we graphed them out as they are written.

When graphed out, it was found that the lower 48 states, Mexico, Central America, the Caribbean, Galapagos, Columbia, Peru, and Ecuador had a pattern of working together. As a result, this region was combined.

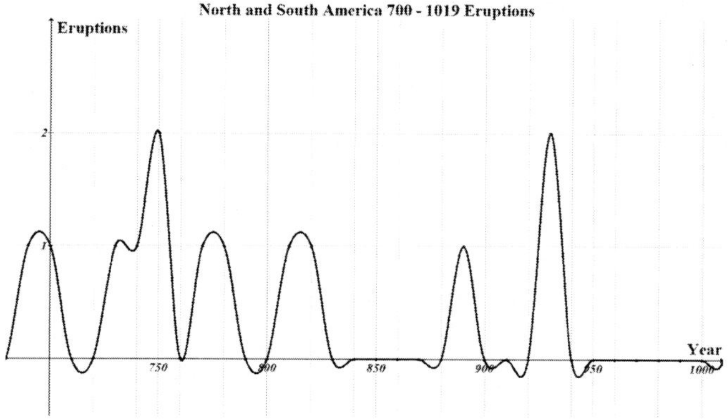

As we look this far back in time, there was a notable grouping of eruptions during this period. Because of the vast numbers of eruptions, this was flagged as our starting point.

In this graph, we can see that there were a total of eleven eruptions during a 300-year period. From ad 730 through ad 829, a 100-year period, eight of these major eruptions took place. This comes to an 80% eruption average per decade. Eruptions from 830 until 1019 lowered to only three over a 190-year period. This comes to an average eruption rate of 15.7% per decade. From 730 to 829, eruptions were more than five times greater than the 190 years that followed.

When this period of time is researched, we find that archeologists acknowledge that the Mayan Empire was at its peak. Shortly after these eruptions, their society collapsed. Recent studies have shown that a cataclysmic breakdown occurred just after these eruptions took place.

Today, archaeology has determined that they had exhausted their resources by stripping the land for the agricultural needs of the vast population and for building and construction. Such events could have easily caused a vast migration of the people to another area with adequate resources. When migrations occur, such as that which occurred during the great Dust Bowl in North America, not everyone leaves. Many communities remain, specifically those that would rely more heavily upon the sea than upon inland agriculture.

The lull period that followed this peak is mirrored in this next graph, and because the following graph is more recent, there is additional credence to the previous era having had an unnatural level of eruptions for this area.

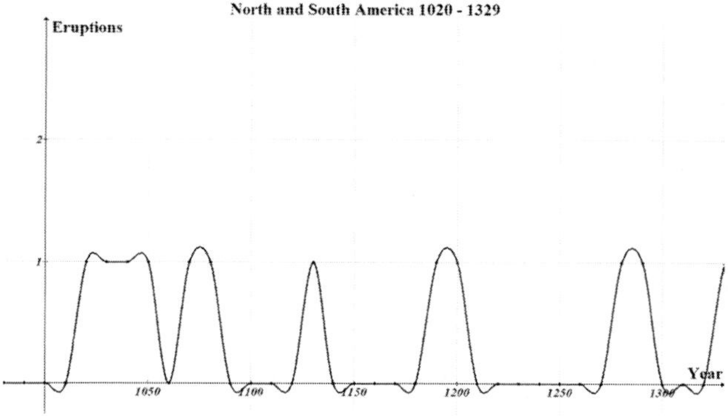

The next period of time where a rise in eruptions began is from 1020 to 1089. After 190 years, from the end of the previous rise in eruptions, we see them elevate again. Beginning in 1020 and then receding by 1089, a 70-year period, six of these eruptions occurred, placing the average eruption rate during this brief period at 86% per decade. Over the remaining 240 years that followed, from 1090 until 1329, there were three eruptions with an average eruption rate of only 13% per decade. This period of time is again greater than five times the eruption rate before, and more than 6.5 times greater after.

A civilization known as the "Anasazi" was known to have existed during this period of time. From 1130 to 1180, there was a 50-year drought that occurred, resulting in the collapse of Chaco Canyon by 1140. Current research has now established that it was during this period of time when the inhabitants vanished. Archeologists have determined that they had wiped out their resources and altered their land for agricultural use and construction material to support a very large-scale population. There is a 200-year time frame (820–1020) from the end of the previous rise in eruptions to the recurrence of these eruptions again. The eruption of 1070 in Arizona was probably a major catalyst, causing many to migrate to new land with better resources. Once again, we see a rise in eruptions, then devastating results.

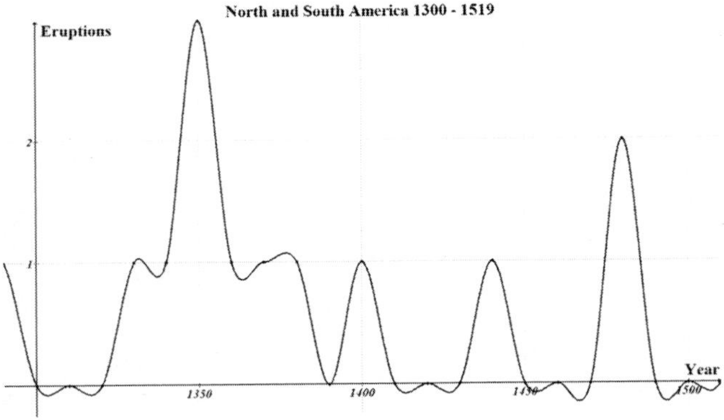

In this graph, there were twelve total eruptions over a 200-year period. From 1330 to 1389, there were eight eruptions in 60 years, an average eruption rate of 133% per decade. This is ten times the eruption rate when compared to the previous years leading up to this period.

From 1390 to 1569, there were eight eruptions over a 180-year period, resulting in an eruption rate of 44% every decade. This is three times the eruption rate from 1330 to

1389 and greater than the two previous lulls in activity. The end of these eruptions also occurred just after the black plague had traveled through both Asia and Europe in the mid-1300s, indicating disease may have altered this civilization. This eruption data indicates that the population may have not evacuated the lands, and disease may have decimated their populations as it did in Europe and Asia.

Both the eruption data and time span mirrors previous civilizations, yet this one appears to be unknown today. Knowing that two large empires emerged during this period of time, it should also appear in a greater level of eruptions later. Out of this era both the Inca and Aztec Empires appear.

Beginning in 1570, there is 100% accuracy to the year of the eruption.

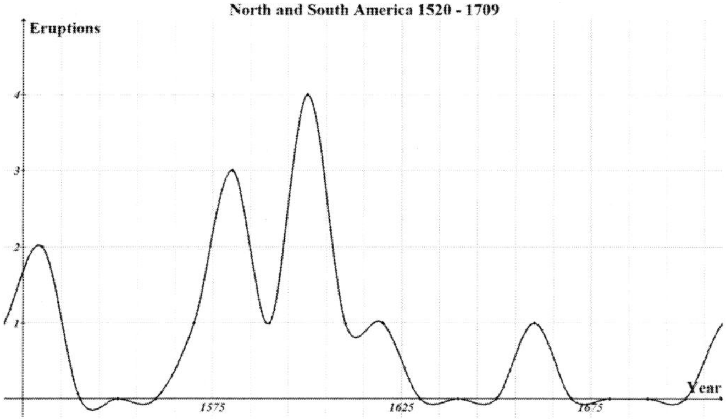

From 1570 to 1619, a 50-year time frame, there were eleven major eruptions. This is an eruption rate of 322% per decade, or more than the period of time, that was already elevated, by more than seven times its eruption activity.

By the 1520s, the Inca and Aztec Empires were wiped out. The years that followed continued to infect the populations throughout North America, having a devastating effect on the population. The Europeans went through and ravaged the land and then left it for the years to come. Although the population had been depleted by the late 1500s, the forests and prairies needed decades to recover. The preliminary heat had already imbedded itself into the plate and the earth responded to man's alteration of the land. This also resulted in a quick reduction in volcanic eruptions shortly thereafter. These events would have had vast impacts on the populations of the people. Eruptions in this region became silent from October 27, 1660, until August 27, 1717, almost 57 years. This was also the period of time when the Maunder Minimum occurred, (1645–1715).

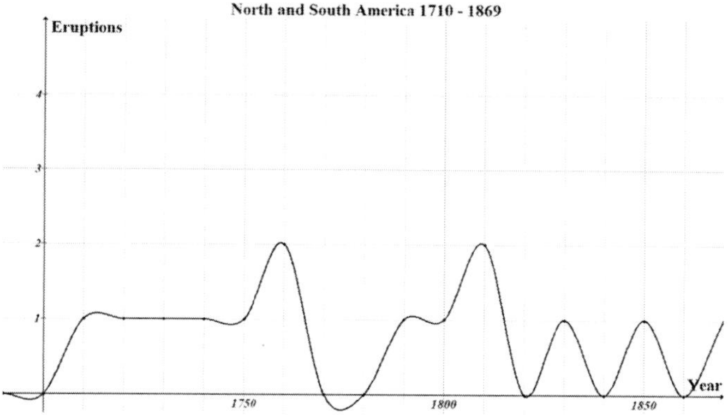

From 1710 to 1869, there were fourteen eruptions over a period of 180 years. This is an average eruption rate of 78% every decade.

America's first colony was established in 1607 in Jamestown, just as the last rise in eruptions was ending. Just over a century later, we begin to see this region slowly rise in eruptions again. There was a lull in activity that preceded the Dalton Minimum (1790–1803) and then a higher than normal, yet stable eruption rate that remained throughout this period of time.

North America was being populated and cleared for its resources and can be seen by the higher than normal eruption rates during this time. Yet, this was also being offset by the massive decline in the population of the Native Americans from war and disease. In 1800, the heart of America remained in close proximity to the oceans, and there still remained a vast wilderness, but changes were occurring rapidly.

With the Louisiana Purchase in 1803, a vast wilderness was ready to be opened up to the people for homesteading, and many would take advantage of this opportunity. As a result, the continent would never look the same again.

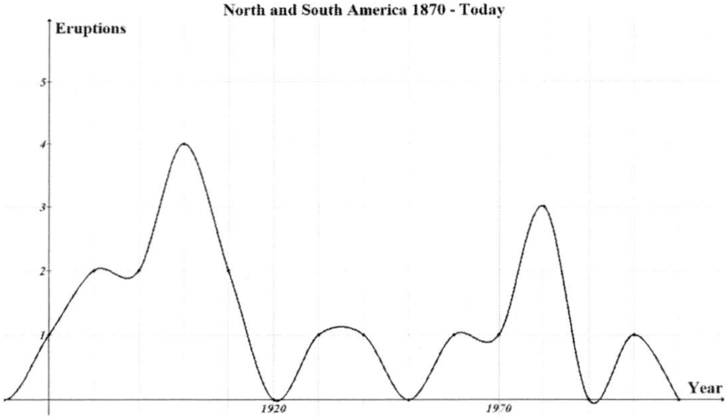

The 1902 eruption of Pelee in the West Indies erupted six days apart and both were categorized as a level-4. Also, in 1982 the eruption of El Chichon in Mexico erupted twice, one week apart, first as a VEI-4 then again as a VEI-5. From a thermodynamic perspective of heat and pressure, such an eruption is significant and is marked as two separate eruptions.

During this 140-year period, we see that there are eighteen eruptions. From 1880 to 1919, there were ten eruptions in a 40-year span. This is a 250% eruption rate per decade. Within 20 years, major land alterations took place causing massive migrations from the central United States. This was then followed by 1920–1979 where the eruption rate declined to four eruptions over a 60-year period. This is an eruption rate of 66% per decade. This is a reduction in these eruptions by 3.8 times their rate and began an era of cooling from 1940 to 1979 due to the vast amounts of air pollutants man was distributing into the atmosphere.

By 1860, 21% of the lower 48 states were cleared for farming and being cleared rapidly. By 1880, the planet was reacting to man altering the land for agricultural and city developments. The United States began to enter the industrial age and the introduction of the railroad, farm tractor, and other machinery allowed us to clear and harvest vast quantities of land rapidly. What once took 200 years or more to convert from forests and prairies into agricultural land was now being accomplished in a few short decades. By the time World War I began in 1917, 44% of the entire land surface of the lower 48 states had been altered for agricultural use to support the war effort. This does not include urban development.

The alterations of the land surface culminated into a mass migration of people from the Midwest shortly after these eruptions took place in the 1930s and left more than

500,000 homeless. By 1940, as it has happened many times in the past, 2.5 million people migrated out of the central United States and took refuge in other areas of the country. This is known today as the great Dust Bowl. After this era, the temperatures cooled as a result of the lack of solar radiance reaching the surface, and this was due to a build up of smog as a result of the industrial age.

The alterations of the land surface in the lower 48 states peaked at 59% in the 1950s, and have declined since for the conversion of farm fields to urban development.

Significant land alterations for agricultural use include the United States, (including Alaska), at 44.88%, 54.94% of Mexico, 75.1% of El Salvador, 43.34% of Nicaragua, 53.86% of Costa Rica, 30% of Panama, 27.96% of Honduras, and 41.66% of Guatemala, 60.28% of Cuba, 47.37% of Jamaica, 61.32% of Haiti, 52.09% of the Dominican Republic, and 21.31% of Puerto Rico according to the World Bank in 2009.[2]

Canada is rated at 7.43% of land use, yet the vast majority of the land is boreal forests and tundra, making it inadequate to support agricultural crops with too short of a growing season. When calculating the amount of arable land usable for agriculture, we find there is 66.7% of usable land is currently being used for farming.[3]

Science today indicates that the primary cause for the great Dust Bowl was improper farming techniques, but due to the cooling that took place after World War II, they never expanded upon their research any further on this occurrence.

Alaska and Russian Peninsula

This is a continual ridge that encompasses the northern Pacific from Alaska through the Kamchatka Peninsula of Russia. This makes up the northwestern edge of the North American plate that is separated by a shallow area of the Pacific Ocean and branches over into Russia. When this area erupts, the plume from the eruption is carried by the jet streams over the North Pacific from Russia and then over Alaska, Arctic, and Northern Canada, increasing the Arctic's intensity, specifically seen during the winter months. In Alaska, the ocean trade winds drive the plume into the Arctic and Northern Canada, increasing their impact on this region.

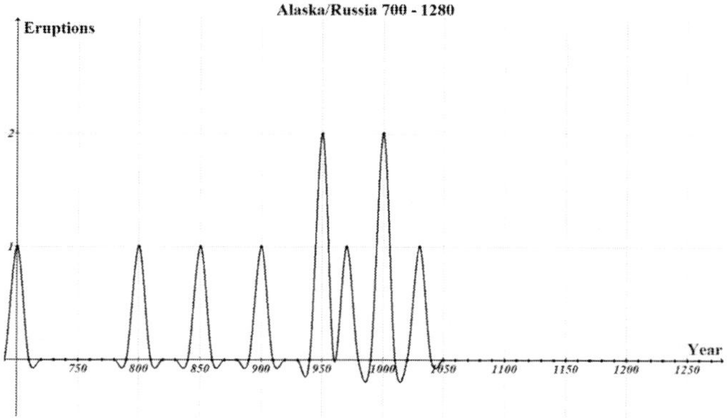

From 750 to 1289, a 540-year period of time, there were nine eruptions and six of these occurred between the years 950 to 1039. Excluding this 90-year period of time, there were three eruptions in a 450-year time span, an average eruption rate of 6% per decade. In the 950s, a rise in eruptions began and continued through the 1030s totaling six eruptions in 90 years. This is an average eruption rate of 66% per decade. This is eleven times the normal eruption rate.

This period of time also coincides with the rise in eruptions in the North American plate from 1020 to 1089. This rise began just prior to the decades of eruptions on the southern section of the North America plate. What is also known is that the Anasazi civilization was further north than the Mayan Empire.

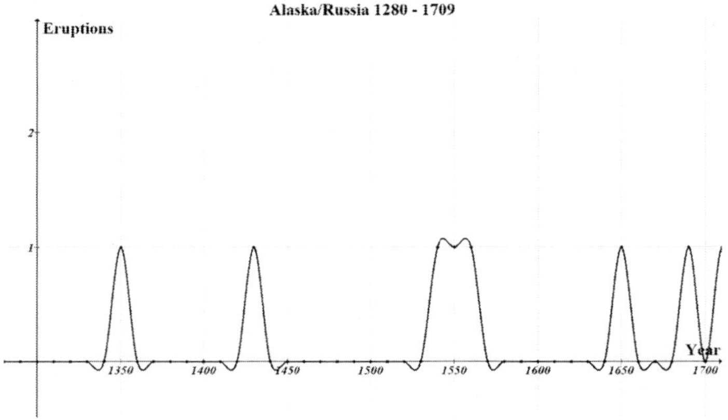

Here, we have a 430-year span of time with seven eruptions. There is a spike in eruptions occurring from 1540 to 1569, a 30-year period. This is a 100% average eruption rate compared to four eruptions in the remaining 400 years or 1% chance of eruption.

The eruptions from 1540 to 1569 appear to be a prelude to the previous research in eruptions when they increased during the periods of 1570 to 1619. This was the period of time that followed the collapse of the Incas and Aztecs.

There were three eruptions during the Maunder Minimum (1645–1715). One eruption occurred in the 1650s, just as it began, one in the 1690s in the middle, and one in 1712, just before it ended. This period of time is the only location where eruptions continued to occur on a regular basis throughout the Maunder Minimum. This time frame corresponds with the colonizing of the North American Continent that began in the early 1600s.

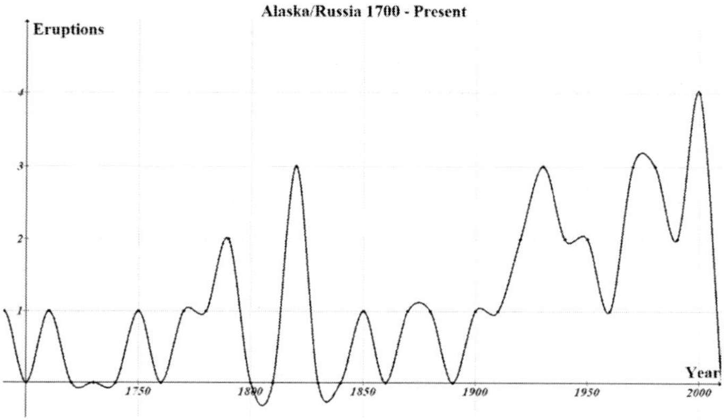

During this 310-year period, we have thirty-five eruptions. This had now risen to a 113% average eruption per decade since North America has been colonized. Since 1920, there have been twenty-one eruptions in 90 years, a 233% chance of an eruption. Since 1970, there have been twelve eruptions in 40 years, a 300% average eruption rate per decade. To put this into perspective, the chances of an eruption every decade went from 1% every decade before the 1700s, 50% in the 1700s, 60% in the 1800s, 190% in the 1900s, and now a 400% chance this past decade 2000–2009.

During the Dalton Minimum (1790–1803), and after a slow and methodical rise, they declined with the last eruption in 1795 then remaining silent until 1825. There was also a noticeable dip in eruptions occurring in North America from 1940 until 1970.

In the early 1800s, there were three major areas in the world that were being altered for agricultural needs of the people in other nations. These were the lower 48 states of the United States, Australia, and Argentina. By the late 1800s, the industrial age was in full bloom and all these areas were being cleared rapidly.

Russia has a vast wilderness with 60% of its land being Taiga, and nearly 11% being Tundra, leaving 29% of the land available.[4] Like Northern Canada, the soil is lacking in nutrients and the seasons are short, cold, and unpredictable. These areas of the world remain primarily in their natural state throughout the world today. Of this 29% remaining, the World Bank rates the agricultural use at 13.16%,[5] or 46.8% of their available fertile land available in Russia today.

Most farmland is found in West Russia and in the southern regions. Other countries that surround this area and their agricultural use of land is Kyrgyzstan with 55.9%, Turkmenistan with 69.4%, Kazakhstan with 77%, Tajikistan with 33.8%, Azerbaijan with 57.6%, Uzbekistan with 62.65%, Georgia with 36.3%, Lithuania with 42.9%, Belarus with 43.9%, Ukraine with 71.3%, Latvia with 29.3%, and Estonia with 21.4%.

Chile

Chile, located in the southern hemisphere, has remained quiet for centuries with only an occasional eruption, until recently. The massive amounts of water that captures, stores, and releases a great amount of energy in this region creates less activity. The eruptions from this region send their plume over Argentina and the South Atlantic, altering the jet streams in the southern hemisphere.

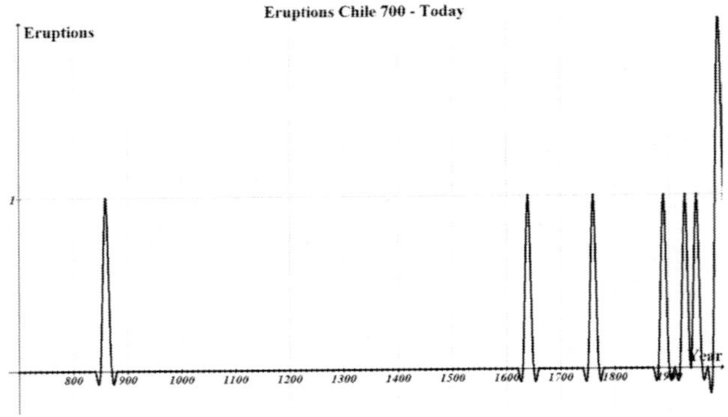

In this graph, we can see the history and infrequent activity of the past eruptions in Chile.

Chile remained quiet through all the previous eruptions that occurred in the north and through both the Maunder and Dalton Minimums. From the 1940s until 1970, there was one eruption during the northern hemisphere's pollution era. This graph indicates that the North American Plate does not affect this region, unlike Columbia, Peru, and Ecuador.

From 700 to 1759, there was only one eruption during this 1060-year period. This is less than a 1% eruption rate per decade. There were only three eruptions that have taken place from 1760–1929, a 170-year span. This is an eruption rate of 2% per decade. From 1930 until 2009, there have been five eruptions in this 80-year period, resulting in a 63% eruption rate per decade.

From the late 1800s and into the early twentieth century, much of the land in Argentina was altered for the agricultural needs, specifically for England after the Revolutionary War. Today, 48.73% of the land in Argentina has been altered for agricultural needs alone according the World Bank,[6] while Brazil is now rated at 31.15%, mainly in the southern and eastern regions. Other South American nations include Paraguay at 51.35%, Uruguay is 83.89%, and Bolivia is 34%. Nations that are encompassed in the Andes Mountains include Chile at 21.19%, Ecuador at 26.77%, and Peru at 16.84%,

Australia

On the eastern edge of the Australian plate, we find New Zealand, and to the northeast, you will find New Guinea, which could be influenced by the much larger Asian plate. Indonesia can be found along the northwest and is directly affected by the Asian and Australian plates.

Due to the population rise in China and its direct effect on the area of Indonesia, we will remove this influence from the graph, detailing a more direct impact man may have had on the Australian plate. This region has 100% dating back to the year 1800.

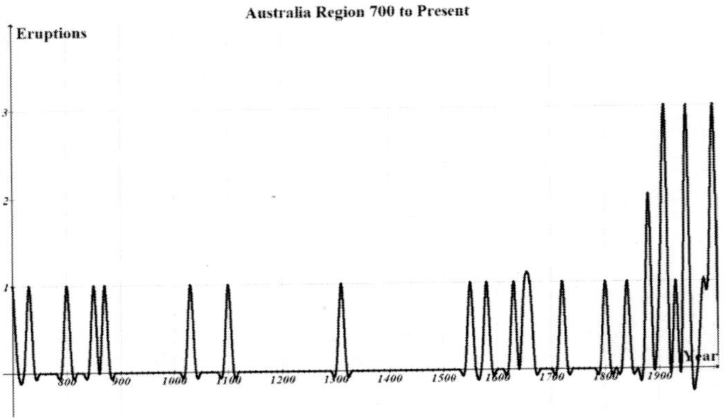

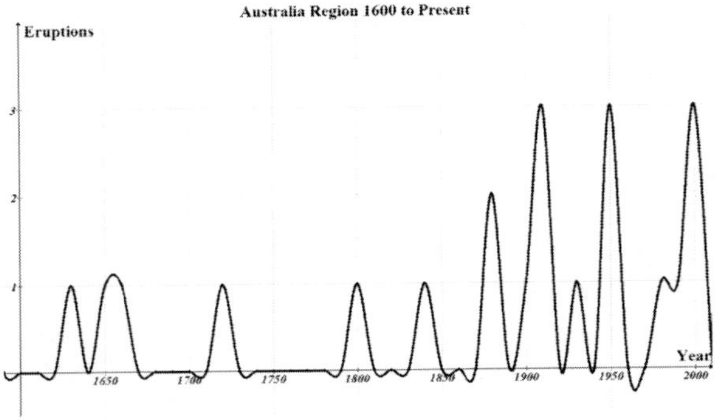

Here, we can see a definable decline in eruptions after February of 1660, shortly after the Maunder Minimum (1645–1715). With the exception of one eruption in 1720, eruptions remained silent until 1800. With one eruption in 1800, a decade after the Dalton Minimum began (1790–

1830), we begin to see a gradual rise beginning in 1840 that carries on to this day.

In 1778, the first settlements began in Australia, and throughout the 1800s, farming increased, primarily by hand. Although there is a rise in eruptions prior to settlement, the New Guinea region on the northern edge of the Australian plate is also buffered by both the Philippine and Asian plates. China's population growth and development directly impacts this region.

During the mid-1800s, the railroads came through and much of the interior was settled and altered for agricultural purposes, specifically for England and Europe's growing population. The Industrial Revolution aided in the transformation of Australia as it did in America, and by 1880, a vast amount of land had been altered for agricultural needs.

During the pollution era of the northern hemisphere, we do find a dip in the 1960s and 1970s, a delay of two decades from the decline in eruptions in the northern hemisphere. The cooling in the northern hemisphere from air pollution would have taken decades to impact the southern hemisphere through the ocean's jet streams and would explain this delayed response.

Today, we see that the increasing population and industrial growth of this area is creating a new form of global dimming in this region through air pollutants. By understanding global dimming in the northern hemisphere, we can conclude that this will, over time, result in a cooling effect on this region by reflecting and reducing the solar radiance upon the oceans causing shifts in the jet streams.

By the 1960s, more than 60% of the entire landmass of Australia was altered from its natural condition to support the world's demand for food. Today, this has declined to support urban development, and Australia reports a 53% use of land for agricultural purposes.[7]

Europe

Volcanic eruptions in Iceland have a great influence on weather over the European and Asian plates. When an eruption occurs, as it did in 2010, the winds and gravitational pull bring the plume over Europe and northern Russia. These eruptions increase the Arctic strength, offering vast cooling over the region and altering the ocean's jet streams. Eruptions in the Mediterranean will cool this large body of water and offer a cooling effect on the entire region, and this plume will flow easterly over the Asian plate. This reduces the midlatitude heat in the region, resulting in a rise in the Arctic's momentum and strength. These alterations can cause massive flooding in some areas, while withholding typical or normal rains from another.

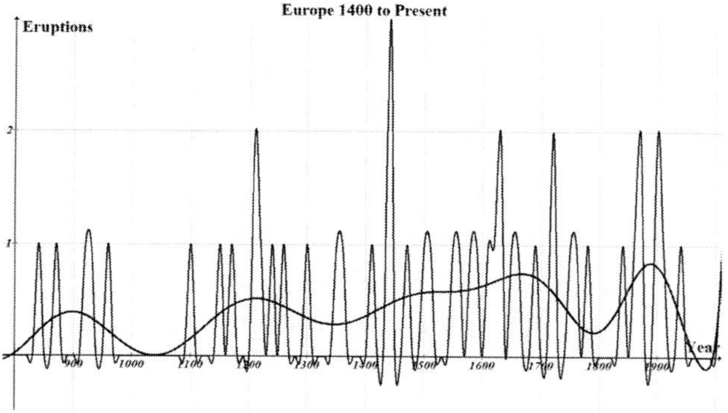

This graph covers the European plate for the last 1,310 years with the exclusion of recent eruptions, 2010 and after. The solid black curve is a 40-year running average of these eruptions. Due to historical data available, there is 100% accuracy to the year of eruptions back to 1550 in this region. Iceland was first settled in 874, making this data far more accurate after this date.

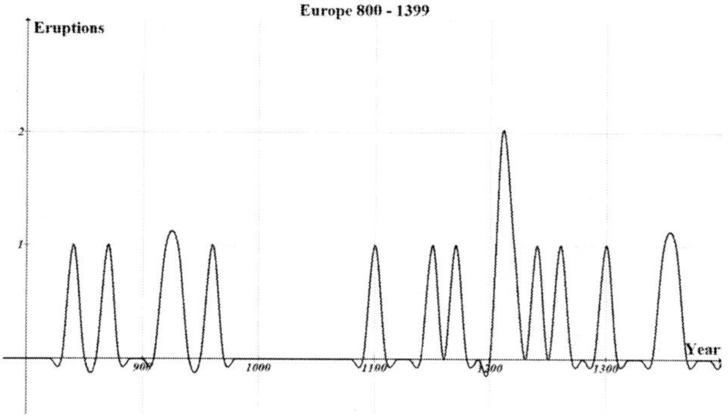

From 800 through 1209, a 410-year period, there were eight major volcanic eruptions around the European plate. This is a 20% eruption rate per decade average. From 1210 to 1269, there were five major eruptions during a 60-year span. This is an 83% eruption rate every decade, more than four times the previous rate. The eruption rates declined from 1270 to 1349 when there was only one eruption in Iceland, an 80-year period, resulting in a decline back to original levels at 12.5% eruption rate per decade. From 1350 to 1549, there were nine eruptions during this 200-year period, a period of increasing but stable eruption rate of 45% per decade.

The time frame from 950 to 1250 is known as the Medieval Warm Period and is associated with the great European expansion when there was massive population growth and an industrial boom in Europe. Much of the resources of the continent during this time were wiped out for agricultural use to supply the necessary resources for the expanding population and construction needs. In addition, because coal was not readily used, wood was used as the primary source of heat during the winters, causing degradation of the land year by year.

Current studies indicate that glacier expansion of the Alps began during this period of time and is making the science community rethink their position on the date in which the "Little Ice Age" began. This period of time had devastating effects on the population, then in the mid-

1300s, the black plague came through and destroyed the populations of both Europe and Asia.

According to known history, it took man 210 years, as this graph indicates, to populate, expand, and deplete the land's resources by hand before resulting in a significant rise in volcanic eruptions.

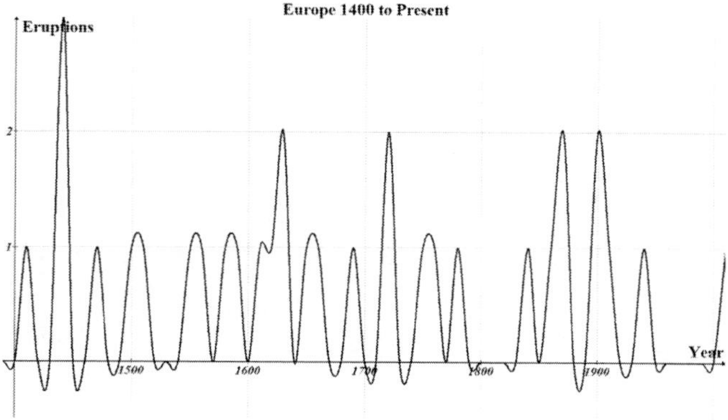

From 1550 to 1669, eruptions began to rise once again as the populations were bursting throughout Europe. During this 120-year period of time, there were ten eruptions, an 83% eruption rate every decade, the same rate as previously experienced in the early-to mid-1200s. There were many problems during this period of time, and this resulted in the great migration to America.

Eruptions began to decline in 1670, just after the Maunder Minimum began (1645–1715) and continued at a slower rate. From 1670 to 1789, a 120-year period, there were six eruptions averaging a 50% eruption rate per decade during this period of time. Although the population continued to rise, it was also offset by the continual migration to the new world. There was a quick emergence in 1720 just after the Maunder Minimum, stabilized again and then were silent from 1790 to 1839 during the Dalton Minimum (1790–1830).

Because Iceland is along the European and American plate, either plate can cause a rise in the eruptions in Iceland, but its proximity to Europe makes Europe's influence greater. Due to the level of alteration of the American plate in the mid-to late-1800s, the American plate may have influenced some of these eruptions. These eruptions peaked during the same period of time as the American plates in 1870 and 1900. All eruptions ended on March 1947. This coincides with World War II when vast bombing took place and the years of pollution that followed, until recently.

By 1975, the EU9—the nine nations in the Europe Union—had altered 53% of their total landmass for agricultural use.[8]

According to the World Bank in 2009, the percentage of land use for agriculture was Germany with 48.6%, France with 53.72%, Spain with 57.44%, Greece with 64.24%, Italy with 47.22%, England with 72.94%, Ireland with 62.07%, Portugal with 38.21%, Romania with 58.92%, Bulgaria with 47.1%, Denmark with 62.76%, Poland 53.17%, Turkey with 51.26%, Croatia 22.28%, Austria, Serbia, Switzerland

39.02%, Czech Republic 55%, Hungry 64.8%, Slovakia with 40.12%, Malta 29.06%, Serbia and Montenegro 57.19%, Luxembourg with 50.58%, Macedonia with 42.31%, Bosnia and Herzegovina with 41.97%, and Albania with 40.84%.

The World Bank includes Taiga and Tundra in their land calculations because life does exist in these regions, but these regions are not suitable for farming. The following nations are Greenland with 0.57%, Norway with 3.39%, Finland with 7.55%, Sweden with 7.64% and Iceland with 22.75%. Like Russia, all arable land suitable for farming is approximately 50% used. The Netherlands with 10% land use for agriculture is built upon low land seawater and channels unsuitable for agricultural use.

Asia

Eruptions along the Asian plate send its plume out over the Pacific Ocean with the trade winds. This can affect the weather greatly in many regions, especially altering jet streams for North America's weather. Any cooling they receive is not from blocking the radiance over their soil as it does in many other regions to the north, but alterations in the ocean's jet streams. This results in alterations to typical rainfalls greater than most regions causing droughts in some areas and extreme flooding in others.

Indonesia is along the southern edge and resides along the equator against the Australian plate and intersects at

New Guinea with the Philippine plate. This is the region plotted in the following graphs.

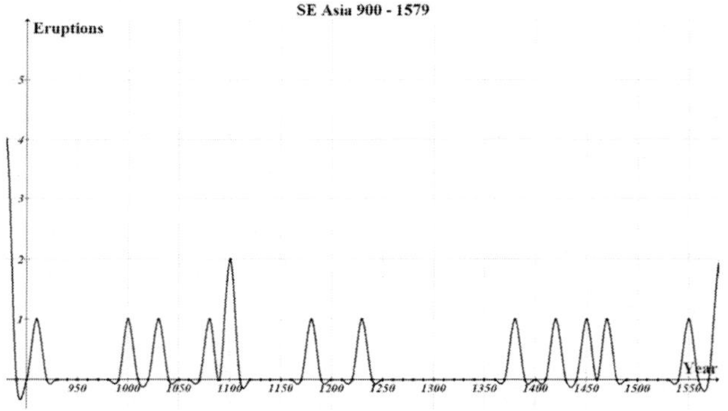

If we compare this with the past history of population growth and previous data, we can see China's population expansion. This would be demonstrated by the depletion of the area's resources increasing volcanic activity.

Of all the volcanic eruptions in this area, only twelve of them have a questionable date from 1000 to 1800, and since 1800, it is 100% accurate. Japan has been one of the most identifiable regions maintaining accurate records dated well before these years.

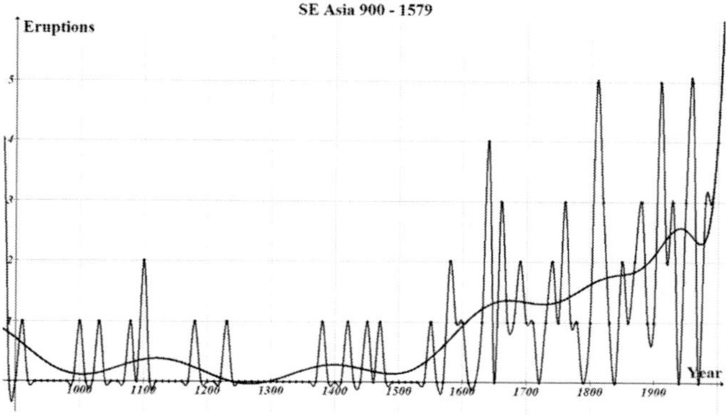

From 900 to 1579, there were thirteen eruptions over this 680-year period, resulting in an eruption rate of 18% per decade. From 900 to 1239, there were eight eruptions in this 340-year period, making this an eruption rate of 24% average per decade. Genghis Khan died in 1227, but was responsible for destroying populations throughout these lands. Then from 1240 to 1379, the period that followed shortly after, there was a clear decline in eruptions and is also associated with the period of time when the Black Death swept across the land affecting not only Europe, but the entire Asian continent as well. It was going to take several decades for the population to rebound from this catastrophic period. From 1380 to 1579, the 200 years that followed, there were five eruptions, resulting in an eruption rate of 25% per decade.

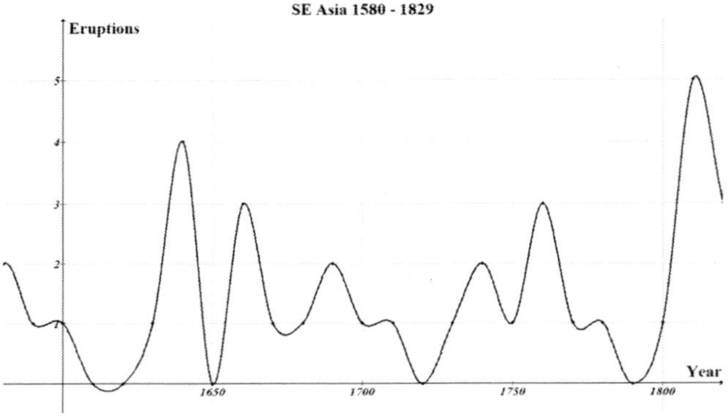

Of these eruptions, only four have a variable eruption date while the remaining are known to the year with 100% accuracy.

From 1580 to 1609, a brief 30-year period, four eruptions occurred, resulting in an eruption rate of 133% per decade. Then from 1610 to 1639, eruptions were dormant in all the regions. From 1640 to 1669, there were seven eruptions in only 30 years, resulting in a rise in eruption rates to 233% per decade for this era. Eruptions slowly declined from 1670 to 1719, a 50-year period, when there were six eruptions resulting in an eruption average of 120% per decade and then became silent through the 1720s. Although the population was rising rapidly during this period, the decline in eruptions corresponds to the Maunder Minimum (1645–1715). After this period, eruptions slowly emerged once again.

From 1730 to 1789, a 60-year period, there were nine eruptions, resulting in an eruption average of 150% per decade. Then in 1790, eruptions became silent with only one in the 1800s. This silence corresponds to the Dalton Minimum from 1790 to 1803. After the Dalton Minimum ended, we see that the area responded with an unprecedented spike of five eruptions in 1810.

China's History of Growth[9]

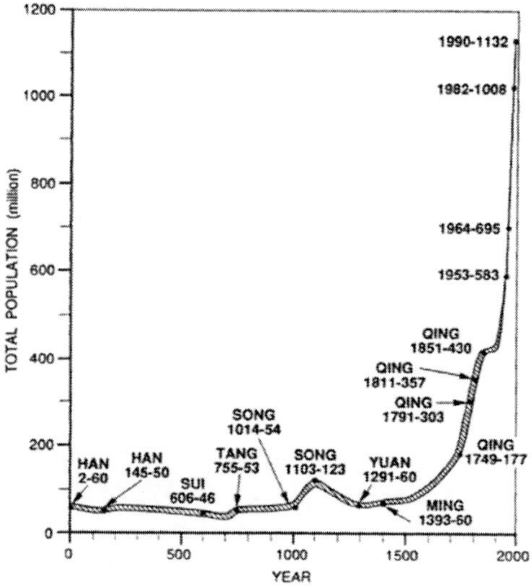

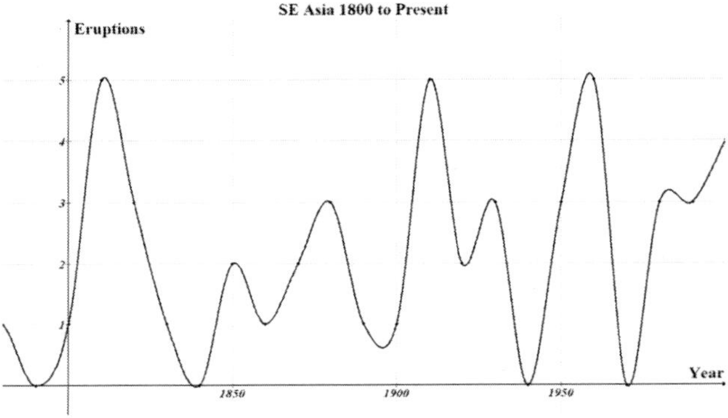

After a spike in eruptions in the 1810s, they began to decline until there was only one eruption in 1831 and no eruptions again until 1853. This is the decade after the Dalton Minimum ended in 1830. We then begin to see an oscillation in an upward momentum as the years progress until the 1930s.

This region was highly active in World War II, and there was a great deal of ordinances and emissions expelled throughout this time. The entire region remained dormant until 1951. We see the eruptions in this area of the world increase until 1966 when the eruptions once again ended, and then began again in 1981. From 1966 until 1975, America engaged in war with Vietnam. Today, it is well understood that the bombing in this region exceeded the tonnage of ordinances dropped during World War II. Europe's silence in eruptions after World War II adds additional credence to this type of war, massive ordinances, suppressing volcanic eruptions. When researching later nuclear testing, there were no signs to indicate they affected eruption activity.

Massive amount of ordinances causes shaking and sifting of the plates. These fissures, located on the edges of the plates are subject to massive shaking causing back fill. It is like an ant farm, and if you shook it, the holes would fill in, similar to a cave collapsing. Over time, the ants would work their way out again, and in this same way, volcanic fissures collapse and fill due to the continual shaking and shifting of the plate.

Since 1950, China and Southeast Asia have more than doubled their populations. In need of increased food supply, this region's resources have been drastically altered to provide food for the people, cities, and developments. They have had 18 eruptions since 1950, or a 266% eruption average per decade today. We have to remember that we are looking at a starting rate of 19% to 24% per decade in its natural state.

Other major nations and their percentages of agriculture use of land in this area include the Philippines with 38.57%, Vietnam 32.48%, Laos 9.22%, Cambodia with 30.9%, Indonesia 26.77%, Thailand with 38.66%, East Timor with 26.09%, Burma with 18.34%, Bangladesh with 69.52%, India 60.51%, Sri Lanka with 36.52%, and Nepal with 29.37%.

Japan

The island of Japan is in the previous calculations and graph, but due to its recent inactivity and its level of accuracy, it is graphed here separately. In this graph, we can go back to 1100 with 100% accuracy to the year, and many to the exact date. Japan is an island on the eastern edge of the Asian plate and is buffered in the north with the western edge of the North American plate, and on the eastern edge is creates the ridge between the Asian and Philippine plates.

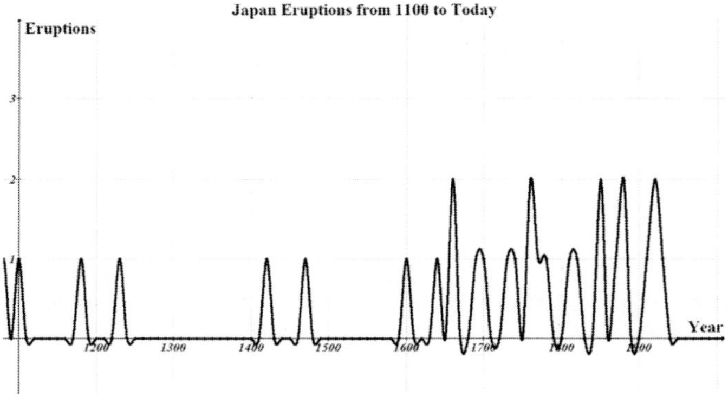

From 1100 through 1600, a 510-year period of time, there were six eruptions. The average eruption rate for this period of time is almost 12% per decade.

From 1600 to 2009, there were twenty-three eruptions during this 400-year period of time. Since the December 24, 1933, eruption, Japan has remained silent. So we can see that these eruptions actually occurred in a 334-year period that results in an eruption average of 65% every decade. This is five times the eruption average prior to this period.

Japan's eruptions grew with the population expansion of mainland Asia, specifically China. After World War II and two nuclear bombs, volcanic eruptions went silent in Japan. The war in Korea that followed World War II and the war in Vietnam through 1975 would have massive impacts. This is another clear indicator of the impacts of man's recent history and wars with large-scale bombings having an impact on eruptions. Korea, being further north, would not have impacts on other regions as it would for Japan. With the current earthquakes experienced in this region, it appears that this area may become active again very soon. One recent sign of activity in this region is the emergence of a new island appearing from the ocean depths off the coast of Japan.

Conclusions

We live on the very thin surface layer of our planet that we call the crust. This crust separates the surface from the very active and dynamic planet beneath us, and there is no greater single event upon the Earth that can alter weather patterns greater than a large volcanic eruption, and they have always been a part of the Earth's cycles. No one can know when or where an eruption will take place, but what we can learn is that there are periods of times when there is a higher or lower risk, but there is always a risk.

When solar cycles rise, frequently a large-scale eruption occurs if the planet warms too much. This in turn offers short-term cooling, but the long-term warming results from expelling elements and gases. This is where carbon dioxide becomes a part of the cycle and aids us in understanding how the planet uses it in its natural form. Rising carbon dioxide levels during solar maximums is a natural response of the planet, but what this does is also allows for a blanket of warmth that stimulates vegetation growth through the solar minimums aiding in stabilizing our atmosphere and surface temperatures. So while the solar cycle declines and outputs decrease, these elements aid in maintaining warmth, like a blanket over the planet. The added carbon dioxide also becomes additional fuel for the vegetation until the solar cycle returns.

So we can conclude that when human activity clears enough area to support its population, an increase in volcanic eruptions occurs. Repeatedly throughout history, when we modify the Earth and a precipice is met, catastrophic results occur. As the human population rises, there is a greater need placed upon the land for food resources. This clearing causes alterations in surface temperatures by exposing the surface to the incoming solar radiation that has a higher heat absorption rate than the natural land cover.

Eruptions declined around the world when there was a decline in solar activity and during our air pollution era of global dimming. At times, even when human activity was rising, eruptions would decline during known solar reduction. This is imperative to understand because this clearly indicates that the planet's actions and responses conform to known physics and comply with our understanding of heating and cooling principles.

Although there is support to the theory that massive eruptions decline during times of large-scale bombing, any attempt to test, alter, or suppress volcanic eruptions will, in time, only result in larger eruptions with far more catastrophic results.

Eruptions: Expansion and Contraction

After I saw this pattern develop, I also noticed another. There was only one eruption on the list in front of me that occurred in December. As I looked further, I found only one other and my thoughts went back to a man who worked at the railroad. He was telling me how the trains were now running on what is called a continuous rail with no seams. He also told me that at the main terminal, they had to allow enough room for expansion and contraction of the tracks because of the massive amount of length they extend during the summer, and how much they shrink in the winter. I decided to look at it with the understanding that all materials, even the earth itself, will expand and contract with temperature changes. This would account for this pattern I was seeing.

With the exception of a few minor elements, all materials follow the laws of expansion and contraction. Even water does, until its tipping point. When elements heat they expand, and when cooled, they contract. When the past 200 years of volcanoes were examined because their exact dates are known, a pattern developed. There were 117 total eruptions, four with unknown dates. This gave us a total of 114 eruptions to examine, and of these, 72 occurred above the tenth parallel north. When taking the Philippines, Russia, Japan, Alaska, the United States, and south to the Caribbean excluding South America, we find that 72% of these volcanoes occur during the first six months of the year, January through June.

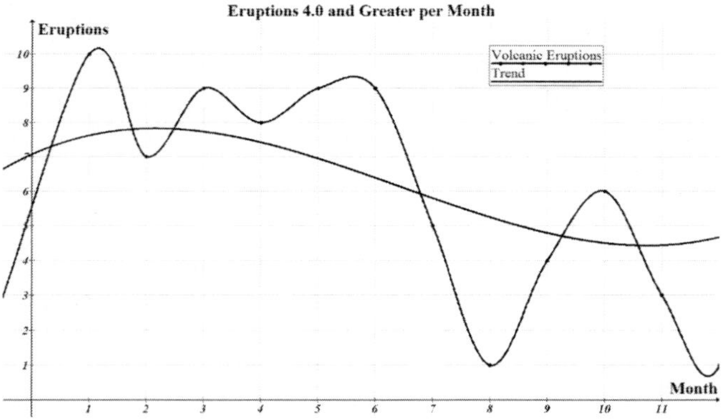

In this graph, we can see a rise in volcanic eruptions that begins in January and settles down after June. There have been 20 of these eruptions that have occurred between

July 1 and December 31. The remaining six months, we have experienced 52 of these eruptions, more than double the eruption rate.

The northern climates of Russia and Alaska expand into July and account for four out of the five noted in this graph for this month. If we exclude these from our calculations, we see that only sixteen out of 72 have occurred during the heated periods of the tectonic plates. This is a 78% potential threat of a large volcanic eruption during the cooling period.

There were no patterns found in the middle and southern latitudes. The southern hemisphere is greatly impacted by the massive oceans, and the middle latitudes maintain a balanced, yet more intensified heat year around. There were 36 total eruptions during this time and 19 occurred from January through June and 17 from July through December. Over the last 200 years, and out of 114 eruptions, only two have occurred in the month of December worldwide. The only periods of time these latitudes were found to have risen were associated with surface alterations and population growth, not with the changes of seasons. This cycle is not coincidental and is expected from a heating and cooling perspective of the expansion and contraction of materials.

The oceans around the earth maintain a continuous and stable temperature year around. Land above the surface and free of permafrost is the only land capable of expanding and contracting as a result.

America is approximately 3,000-miles wide (15,840,000 feet). Granite, limestone, and marble all have an expansion rate of .0000044 inch per degree.[10] Depending upon your location, the topsoil can alter by over 60 degrees during the change of seasons, and the air temperatures much greater. If the North American plate were one solid piece of granite 3,000 miles wide, the plate would expand or contract 84 feet for every degree in alteration. This is 42 feet both east and west. This means the North American plate, and all plates will, according to the laws, apply more pressure outwards and against other plates during times of warmth, and less pressure along these regions when cooled, explaining this oscillation in volcanic activity.

Granite makes up only 25% of the tectonic plates and for its size has the smallest expansion properties of the natural elements along with limestone and marble, meaning the expansion will be greater due to water, clay, and sand. The heat increases molecular movement that also increases pressure within a sealed system. As the planet tilts on its axis and enters into summer, it heats and the plates expand. By July, they are fully expanded in the southern region of the northern hemisphere and expands northward until the northern region is fully expanded by August. The opposite would apply to the southern hemisphere.

The majority of these volcanoes are located along the edges of the tectonic plates where one plate meets up with another. The most active is the area in the Pacific that is known today as the Ring of Fire. As the plates heat, they expand and apply pressure along the contact points, pinching them off and allowing the pressure to build. This is a vital function, shutting down these fissures and allowing the pressure to rise within the core. If pressure were released during the summer when the planet heated, the following winter would cause contraction, resulting in severe and catastrophic earthquakes.

The force at work is the same force you would find if you fill a bottle with hot water, seal it, and then cool it. The contraction in a sealed container creates a vacuum and the outer shell will collapse in on itself, and the same would happen to the planet if this pressure were not sealed and contained during the heating periods of summer. From a thermodynamic perspective, there is reason and understanding to this because we use this principle in our cars. Our radiator cap seals our coolant in under pressure and allows the boiling point to rise as a result, when cooled it is allowed to vent.

The heat beneath the Earth's plates rising up from the core alters slightly during seasonal cycles. Here in Minnesota, this temperature variable can be seen and measured six feet underground by monitoring the water temperatures early in the morning when water has been dormant in the line. As each day, season, and solar cycle alters, the heat slowly rises and lowers, causing expansion and contraction twice every year, much in the same way that we take a breath. These cycles allow for slow methodical alterations due to outside influences such as changes in solar activity, massive forest or prairie fires, freezing temperatures, or even meteor impacts.

If the alterations are natural, such as a strong solar output during a cycle, the planet responds and then settles down. What has been seen throughout this research is when mankind's population rises to a level where the land's natural resources are cleared and exposed to the solar radiance, large-scale eruptions increase in occurrences. When the land is not allowed to return to its natural state and man continues to attempt to work the land, Mother Nature gradually turns violent over the following decades. The plates heat, expand, and apply pressure on neighboring plates acting as a shut off valve for further natural eruptions. The heat is then forced to build beneath the surface slowly rising year by year. This requires decades of increased radiance applied on the surface due to the planet's axial shift constantly altering this heat. In time, these effects are experienced upon the surface in the form of droughts, intense storms, wildfires, and is often accompanied by massive flooding. Increased evaporation is a by-product of increased heat and leads to an increase in precipitation and storms. In the planet's history, this has always meant migration from the affected area, but with technology available today, there have been many changes. Such innovations such as air-conditioning and heating, available freshwater, and increased farming technology have changed the face of this world. Many areas today are able to produce crops from fields that were often unproductive.

The middle latitudes are more susceptible to surface alterations due to the amount and intensity of the sun's radiance on these areas. This makes these areas more vulnerable to faster responses due to the alterations of the surface than latitudes that are closer to the poles. As the tropical forests dwindle away, heating becomes more rapid and is evident by the rise in eruptions in Chile today. The forest offers a damp, cool, and shaded environment that is lost when converted.

As the plates cool through the fall and into the winter months, they reach their maximum contraction stage from the lack of solar energy and then relieve pressure along the plate borders. If the internal pressure is high enough, and the plates contract enough to decrease the pressure along these points, eruptions rise in occurrence.

Heat generated from the core expands under the plates and aids in their movements around the core. Driven by the internal core's heat, pressure, and fluid movement, along with the gravitational pull and ocean tides, the plates slowly crawl across the surface. As each season comes and goes, like a snail, they expand and then contract slowly moving across the planet's surface.

If the world was reversed and the water mass was in the north and the landmasses were in the south, there would be catastrophic effects. Due to the heat, pressure, and expansion, large-scale eruptions would occur, as it once did in Earth's history. Science indicates that early on, the landmasses were connected and located near the equator and that extremely large-scale eruptions occurred early on. By known physics and the data accumulated, the landmasses would have been blown apart by eruptions and massive earthquakes, resulting in faster acceleration of tectonic movement early on in Earth's history and supports this theory.

The planet distributed the land and water to its current position as part of its nature to utilize the water, optimize its cooling, stabilize itself, and achieve balance. This was all a part of the early development of our planet to become the way it is today. This heat and pressure building up today could lead to potentially very large-scale eruptions and earthquakes of large magnitudes as it did early in its development if balance is not obtained. Human activity continues to alter and jeopardizes this balance.

Eruptions and Their Impact

Eruptions around the world repeatedly demonstrated a decline either during, or shortly after the Maunder and Dalton Minimums. These periods of time were also associated with known climate cooling indicating a direct link between these two activities—solar activity and eruptions.

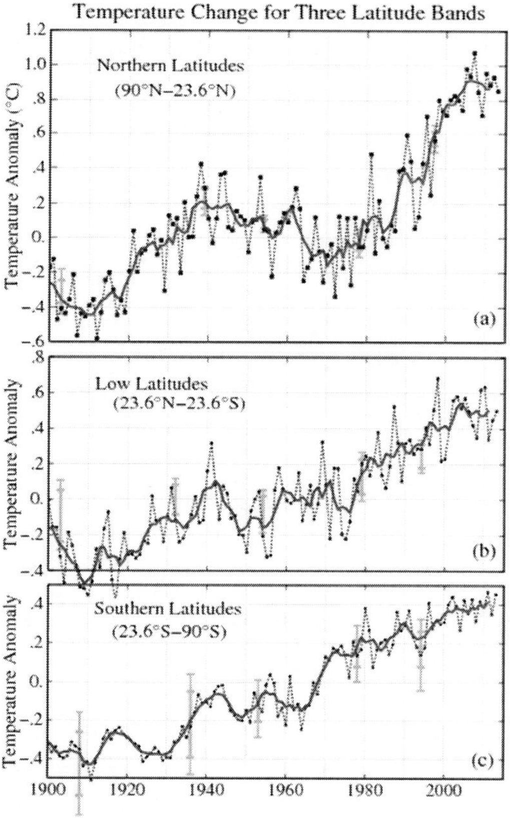

Currently, the northern hemisphere is experiencing faster warming trends than Antarctica is in the south as this graph indicates. We know there are two variables that exist between the two hemispheres: population and land distribution.

The cooling effect during the pollution era was primarily confined in the northern hemisphere as these graphs indicate.[11] The resistance to incoming solar radiance remained in a lateral direction around the planet and did not alter the other hemisphere.

The location of an eruption has a much greater effect along the latitude of the eruption as well as the distance from the eruption. An eruption, for example, in New Zealand, has little to no effect on Russia in the north, and an eruption in Alaska has little to no effect on the people in Australia in the south. The initial blast affects the entire area of the planet where it erupts and is proportionate to its size and type of eruption. The solar winds mix this debris from west to east direction through the same gravitational forces that move our tides and inland weather from west to east and slowly dispense it throughout the atmosphere. On the lower levels, the ocean's jet streams drive its direction and impact region through the trade winds.

In 1991, Mt. Pinatubo erupted at a level six in the Philippines. The science community has emphasized the cooling effect of this one event, but statistics and historical data reveal another. Although this was a very large eruption causing massive effects on our weather, the following impacts need to be included.

Beginning in 1990, prior to this eruption, a VEI-4 erupted on January 30, 1990, in the Kamchatka Peninsula of Russia and another in Indonesia later that year on

June 15, 1990. The winds from the Kamchatka eruption would have a direct impact over the northern Pacific and the North American plate, increasing the Arctic strength over this region affecting Minnesota's following winter of 1991–1992. The eruption in Indonesia would have a direct effect on their region by altering the temperate waters of the Pacific and Indian oceans. Then in 1991, the Kuwaiti oil fires began covering much of the Asian plate in a cloud of smoke, obscuring a very large portion of their land from the solar radiance throughout the year. With the understanding of the effects of shading the planet during the air pollution era, such a vast amount of smoke would have a great impact on this region being the largest exposed tectonic plate in the world. The amount and levels of smoke was obvious by the satellite images that were shown to the public during this time of the Iraqi withdrawal from Kuwait.

All of this occurred prior to the Mt. Pinatubo eruption in 1991. This eruption would reduce the radiance and alter the jet streams of the world's largest body of water, the Pacific Ocean. The Pacific Ocean, as the world's largest ocean, also retains the largest amount of energy. When this massive energy held within of the Pacific Ocean declines, it also takes time to rebuild, resulting in a multiyear cooling effect.

Just after this eruption, a VEI-5 erupted on August 12 in Chile, just two months later. The winds and jet streams would move this out over Argentina and the southern Atlantic, offering vast cooling to these waters.

Neither of these eruptions seriously altered the Arctic's influence in Minnesota's winter as other volcanoes do that are closer in proximity as demonstrated by the following graph. It's the accumulation of all these effects and no one specific event that resulted in worldwide cooling of the planet, specifically the summer of 1992. A few years later, we begin to see a slow and methodical rise in temperatures again.

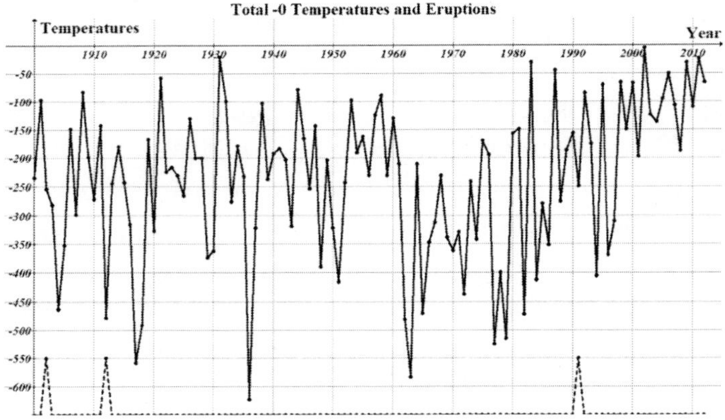

The impact of eruptions becomes evident in this graph showing level six eruptions and the impacts it has had on our winter months here in Minnesota since 1900. The greater number of below zero temperatures for a year, the greater the influence the region has from either the increased cooling effect in the north, or a declined heating effect from the Gulf Stream.

In Santa Maria, Guatemala, on October 24, 1902, a VEI-6 erupted that had a large impact on our weather systems until 1904. In June of 1912, a level six erupted in the Alaska Peninsula that demonstrated a serious impact on our winter. This was followed by a level five eruption in Mexico the following year in 1913 that had no apparent effect.

In comparison, when we reflect upon the June of 1991 eruption of Mt. Pinatubo, we see there is no immediate and direct effect, especially when we consider the level five eruption that occurred in Chile the following month. Eruptions, like nuclear bombs, have the greatest impact near the eruption. The impacts are carried by the winds and dissipate as you travel away from the location. Altering the atmospheric and oceanic jet streams is the major way eruptions impact our weather.

Understanding Recent Eruptions

If we apply this knowledge into understanding recent eruptions, we can look back at the past few years and see how they have impacted our planet's weather patterns.

In 2009, Asia and Europe were experiencing rising warmth and on June 11, 2009, the planet responded to this heat with a VEI-4 eruption in the Kuril Islands of Russia. In Minnesota, we recall the winter following this year was a cooler than usual winter with plentiful snow that seemed more normal. This eruption had very little influence on Asia and Europe.

In 2010, the heat intensified in Russia and was known as "The year Russia burned." Massive deaths were reported as a result of the heat throughout Europe and Asia. On April 14, a VEI-4 eruption occurred in Iceland that closed airports throughout Europe. This eruption occurred too late to have an immediate impact on the region but was felt by the following winter over much of Europe and Asia as the temperatures began to decline. There is always a delay between applied heat or cooling and the air temperatures measured. In North America, the heat began to return, overcoming the 2009 eruption, and in 2010, there was an excellent year for crop production.

On November 9, 2010, an eruption occurred in Indonesia causing massive influx in the ocean temperatures of the region. This cooling over the ocean between the Pacific and Indian oceans caused massive flooding and droughts in various regions throughout Southeast Asia and Australia, devastating crop productions through 2011. This had no measurable impact on North America, Northern Asia or Europe as the heat continued to build in these regions, causing massive crop failures.

After 2010, these large eruptions around the planet have become silent, and throughout the world, the heat began to rise as the solar cycle became active. Both 2011 and 2012 were years of poor crop production around the world, and were progressively worsening. Today, in 2014, we begin to see the decline in solar activity, and with it, the potential for large-scale volcanic eruptions rise drastically.

There was a spike in eruptions during the early 1900s, and within 30 years, the planet was responding by bringing in the great Dust Bowl. Our last spike was in the 1980s, bringing us to within the timelines, and from what the data indicates, our weather patterns are returning to the great Dust Bowl era.

Volcanoes of this magnitude have a relatively short-term cooling effect that follows after they erupt. They extend their cooling effects by altering ocean temperatures and jet streams. Like a pressure cooker, the more heat, the more eruptions we can expect. If the planet continues to heat beyond its natural levels, the plates permanently expand, shutting down the planet's eruptions and then notable alterations of the planet's surface follows. After the heat subsides, the carbon dioxide and other elements that are released affect the planet through the solar minimum by maintaining its warmth and fueling the vegetation. This greenhouse effect will increase the warmth, rainfall, and vegetation. The vegetation grows and consumes this excess carbon dioxide until the next solar cycle in a natural environment. The additional carbon dioxide being injected into an atmosphere that is already saturated will have a reverberating effect, increasing the warmth during the solar minimums, as it did in the 2000s. Understanding that a rise in the Earth's heat will increase volcanic activity and wildfires, and that both are natural occurrences and add to carbon dioxide levels brought me to the next question, how exactly does carbon dioxide work within our planet?

LAND ALTERATIONS AND THEIR EFFECTS

Vegetation

As I look down from an airplane and drive across the country early in the spring, I see vast amounts of barren land in comparison to the thriving green forests and prairies that once engulfed our landscape. This reminds me of a massive wildfire that encompasses the entire surface, or the aftermath of a meteor impact striking the planet and wiping out the surface vegetation every spring and I have to ask, "What's the impact of this activity?"

We all learned in school that vegetation takes in carbon dioxide and releases oxygen and that we take in oxygen and release carbon dioxide. There is harmony that exists in this relationship and this is what is explored, this transitional effect. When we look to this vegetation, there are four basic needs—namely, adequate sunlight, air (specifically carbon dioxide), water, and soil nutrients.

Carbon Dioxide

Carbon dioxide is taken in during the process of photosynthesis, and this process only occurs during the hours of sunlight when the temperatures allow for it. As the vegetation processes this carbon dioxide, it grows and applies this into its leaves, trunk, stems, and roots. Newly developed ecosystems (for example, after a wildfire) sequester vast amounts of carbon dioxide necessary for new growth in comparison to older forests where we find slower growth and decay that releases carbon dioxide, decreasing its impact. The faster the growth of an ecosystem, the more carbon dioxide is demanded.

The carbon dioxide cycle is more complex and intertwined with life than we understand. One image that comes to my mind is to see the wildebeest consuming a prairie and leaving it barren, to only find that within a very short period, the fields have returned. This would have been much like North America and the bison in the early 1800s. In harmony, the wildebeests consumed vast amounts of vegetation and use it for energy and growth. What is not used is then released as they shed and leave waste behind. They also release carbon dioxide back into the atmosphere through their breathing while utilizing the oxygen that the plants release during their photosynthetic process. Consuming this vegetation prompts vast growth, and it's in this growth where we find vast amounts of carbon dioxide being sequestered from the atmosphere. Insects are even more vital as they consume even more vegetation than mammals. New and widespread growth, specifically in the spring, captures vast amounts of carbon dioxide. When man alters the land surface into farmland and city developments, he also alters the atmospheric carbon dioxide concentrations due to the decline in the planet's natural vegetation and alterations in this cycle.

Carbon dioxide is one of the elements in the atmosphere that increases warmth in an environment by resisting outgoing heat. Science was debating this issue prior to, and during, the great Dust Bowl. World War II began and the planet began to cool, silencing the critics and science community. Over the recent years, this argument has resurfaced once again and remains in question by many. Because the carbon dioxide levels were rising while temperatures declined during the 1950s and 1960s, this clearly indicates that the effects of carbon dioxide on our temperatures are dwarfed in comparison to altering the planet's surface. Altering the surface is also one of the major initial causes behind rising carbon dioxide levels, but the exposure has and continues to drive our climate changes. Any argument for or against carbon dioxide must include this within their calculations. Carbon dioxide levels do not cause heating but is one of the elements that aids in controlling it, and this is why there are so many scenarios and long-term climate speculation.

In its natural form, the vegetation around the world alters the atmospheric concentration levels of carbon dioxide and is the only process that can both raise and lower these levels. This is determined by many different conditions but massive droughts and wildfires are the most common event that will naturally alter these levels. Although life, eruptions, and other events will raise carbon dioxide concentration levels, only vegetation can lower them. This is the key to understanding carbon dioxide.

Carbon dioxide levels are altered through means that go much further than emissions, but our emissions are one of many alterations man has created that are accelerating our problems today. Understanding these cycles and how they impact our climate is necessary in order to determine the degree of impact upon the planet's heating and cooling.

Using the statistics provided by the Intergovernmental Panel on Climate Change, I used their data to create a comparison chart for understanding the impacts of farmland and urban development on the world's carbon dioxide levels. Here are their estimated carbon stocks that are found in various environments around the planet.[1]

Table 1-1: Global carbon stocks in vegetation and top 1 m of soils (based on WBGU, 1998).

Biome	Area (10^6 km²)	Carbon Stocks (Gt C)		
		Vegetation	*Soils*	*Total*
Tropical forests	17.6	212	216	428
Temperate forests	10.4	59	100	159
Boreal forests	13.7	88	471	559
Tropical savannas	22.5	66	264	330
Temperate grasslands	12.5	9	295	304
Deserts and semideserts	45.5	8	191	199
Tundra	9.5	6	121	127
Wetlands	3.5	15	225	240
Croplands	16.0	3	128	131
Total	151.2	466	2011	2477

Using this graph of carbon stocks for different environments, we can convert the areas into equal proportions to demonstrate the differences between these various environments. Here is a cross comparison of carbon stocks with equal area size, demonstrating the degree of alterations that occur by converting one biome to another.

Table 1-1: Global carbon stocks in vegetation and top 1 m of soils (based on WBGU, 1998).

Biome	Area (10^6 km^2)	Carbon Stocks (Gt C)		
		Vegetation	*Soils*	*Total*
Tropical forests	10	120	123	243
Temperate forests	10	57	96	153
Boreal forests	10	64	344	408
Tropical savannas	10	29	117	146
Temperate grasslands	10	7	236	243
Deserts and semideserts	10	2	42	44
Tundra	10	6	127	133
Wetlands	10	43	643	686
Croplands	10	2	80	82

This chart represents carbon stocks based upon equal sizes of land, so we can understand the impact of altering nature. Cropland contains more carbon on the surface through their roots than semideserts and deserts, but above the soil they have an equal amount.

When we converted the forests and prairies into croplands, we created desert environments across the land with various long-term impacts including the water tables and the carbon cycle. Additionally, we altered and channeled the water to maximize the area of land for farming that resulted in the loss of large amounts of wetlands. In the tar sands of Northern Canada and vast wilderness of Russia, we find the Boreal Forest region. As we convert this land for oil and other resources, we can see the impacts this will have on the carbon stockpiles for these regions as massive amounts are in the soils.

As we can see, the carbon cycle has a two-part structure, just like our checkbooks. There are both deposits (sequestering) and checks (emissions) that need to be considered. If both sides are not accounted for, then a true balance can never be attained.

Science today doesn't calculate the transformation of land into any of their figures for rising carbon dioxide levels and only includes the use of fossil fuels. In doing so, they mislead themselves and others by disregarding this severe and widespread impact on the world's carbon dioxide levels. Evidence is demonstrated by the early rise in carbon dioxide levels rising prior to 1800.

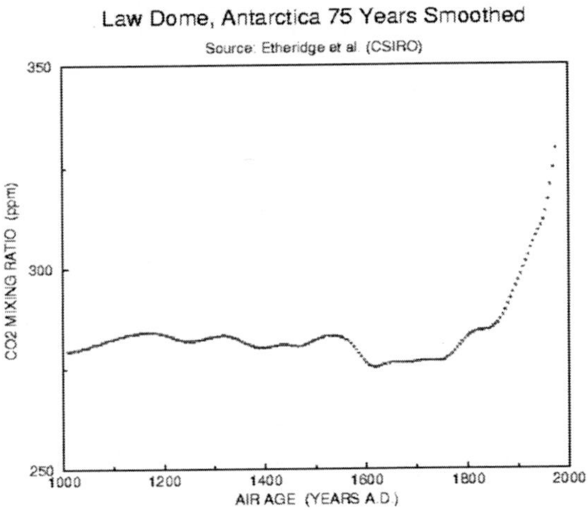

In this graph, we can see the carbon dioxide levels extracted from ice cores over the last 1,000 years.[2] Nearly half of our current rise in carbon dioxide levels can be attributed to these vast alterations of land, the other half is due to emissions.

If we follow man's activity in this carbon dioxide graph taken from Antarctica, we can begin to see the carbon dioxide levels slowly rising during the European warming period from ad 1000–1200, until the volcanic eruptions began in Iceland. The Anasazi collapsed during this time frame in North America and the little ice age began in Europe. So as the populations around the globe reduced, the planet began to slowly recover; the vegetation levels increased, resulting in a reduction in world carbon dioxide levels.

Population growth and carbon dioxide levels began to rise again until the mid-1300s when the black plague wiped out a vast amount of the population. Over the following years, the land returned to its natural state and carbon dioxide concentration levels once again began to decline proportionate to vegetation growth.

Soon, the population began to grow in both North America and Europe beyond previous levels and the carbon dioxide levels began to rise in response to the declining resources. After the populations of the indigenous people of North America were decimated by war and diseases in the early 1500s brought over from Europe, the land returned to its natural state. As a result of this massive decline in North American population and wars in Europe, the planet began to return to its natural state, and the North American continent would remain virtually untouched for the next 200 years and is demonstrated by a period of stable and lower carbon dioxide levels. After the Revolutionary War and the Louisiana Purchase in 1803, the North American continent would never look the same again. Adding additional carbon dioxide into the atmosphere today is demonstrated by the sharp increase we see at the end of this graph.

This clearly indicates that land alterations do alter atmospheric carbon dioxide concentrations and must be included in any analysis of carbon dioxide levels. What this means is that stopping *all* emissions today will not stop the carbon dioxide levels from rising. An extensive widespread growing plan in conjunction with a reduction in emissions is necessary to reverse our current trends.

The Kyoto Protocol requires carbon tax on only emissions, but not addressing a country's level of land alteration into the formula for taxation is fruitless, and formulating a balanced carbon dioxide level is impossible. Farming increases carbon dioxide outputs and decreases the inputs, meaning those counties that would be charged higher taxes are also those nations supplying much of the planet's food resources. The final result will only be rising prices due to additional taxation and the financial resources used from this tax will not resolve our problems. Without growing the planet on a massive scale to sequester the carbon dioxide, any attempt to resolve our rising temperatures will end in failure.

As forests, prairies, and jungles give way to farmland and urban developments, we also find vast alteration in the carbon dioxide cycles. If these alterations continue to remove vegetation, then a continual rise in carbon dioxide levels is expected. It's a natural by-product of these changes. The carbon dioxide graphs created by the science communities today do not adjust for this variable and apply all increases to fossil fuels.

I had determined that the planet was warming and the cause was human activity but not carbon dioxide from fossil fuels as science suggests, although it is a factor.

The primary cause is the depletion of the planet's natural resources, resulting in a loss of resistance to the incoming solar radiation, which also accounts for the constant rise in carbon dioxide levels. Use of fossil fuels explains the rapid rise since World War II. Exactly how this occurs and is accomplished is what needed to be explored further. The decrease in surface vegetation is an element that has known effects on the carbon dioxide budget in the atmosphere and would result in rising carbon dioxide levels. Although the temperature variables and thermodynamic principles support this theory, until definitive evidence could be provided, this was only a theory subject to scrutiny, the same as the carbon dioxide argument is today.

Over the following couple of years, I researched a vast array of areas looking for patterns and cycles along with any historical evidence to support this theory. For three years, I researched different avenues and even with the vast amount of data collected to support this concept, no concrete evidence emerged. Without evidence to support this theory, it was only speculation.

Coolant and Circulation: Water

Within any machine, coolant and circulation is a necessary process to maintain stable operation, no matter what the external environment may be. Four-fifths of our planet is water so as the Earth heats, water evaporates and takes to the atmosphere where it rises, cools, and condenses. Then, when saturated, returns back to the Earth through the force of gravity in the form of precipitation. In this process, there is both cloud cover and moisture returning back to the planet in a continuous cycle that increases with heat. This offers both shade and life-sustaining purified rains to cool the surface. The more heat, the more evaporation. Therefore, the more water is released back upon the Earth. This process is self-generated and self-correcting acting as a thermostat, water pump, radiator, and purifier for the planet.

When discussing water, many variables need to be considered. The amount available for storage, the amount going in through precipitation and what is departing through evaporation. All of these have multiple variables that can either increase or decrease these amounts, and this is what is researched. Too much cooling can prevent a machine from reaching optimum operating temperature, and too little can cause catastrophic overheating. In the end, over thousands of years, the Earth has created a harmonic level, and altering water levels will alter surface climate conditions.

Water is an element that works the opposite of nearly all other compounds on the planet. When we think of heat, we think of expansion. When we think of cooling, we think of contraction. With water, the opposite is true. When we fill the ice tray to the top, we soon find the ice overflowing, demonstrating an expansion when cooled. As it turns to a solid, it also becomes lighter and floats. As water heats, there is increased expansion until the evaporation point is reached, then it contracts into a vapor and these principles are vital to our planet's cooling. Water is the only element on our planet that naturally exists in all three states as a solid, liquid, and gas.

The planet's daily rotation, this heating and cooling effect from day to night, maintains the balance of water on the surface in combination with the atmosphere. The hotter it becomes, the more water will turn to vapor, saturating the atmosphere; acting as a biological thermostat and making adjustments.

Water is a vital part of our planet that is not only a major part of our life, but also one of our basic needs. Water also makes up the greatest portion of our bodies and is continually flushing and replenishing our bodies with necessary fluids. All the water within this world has been here since nearly the beginning of the formation of our planet and nearly all of it remains either in, on, or above the earth in the lower portion of the atmosphere. This water that we drink and use has been through life before, purified, and recycled over and over. Just as we use water for washing our clothes or cleaning the baby's bottle, so does the planet use water for purification. As it evaporates, it rises up in a purified state, cooling as it rises. As the water comes down in the form of precipitation, it brings with it contaminants to the surface. As it works its way through the soils, the planet purifies it again and adds life-sustaining minerals to it.

Water is also used for heating and cooling due to its unique qualities. We use water to cool our cars and nuclear power plants, and to transmit heat throughout a building. Just as we use it in many applications, so does the Earth. Our planet's surface is estimated to be 70% water, and the oceans act as large cooler for a very large reactor within our planet that we call the core. Water's unique qualities allow it to not only absorb heat rapidly, but can also dissipate it quickly too. Water creates a large buffer against alterations in solar energy both absorbing and dissipating this heat daily.

Water also has a quality they call sublimation. This is the process of turning from a solid (ice) into a gas (water vapor) bypassing the fluid state. The opposite is also true. Water vapor can turn from a gas to a solid state bypassing the liquid state and this is called desublimation or deposition.

As water evaporates, it is warmer than the surrounding air so it begins to rise. When it reaches its freezing point, it crystallizes and gravity takes effect, bringing it back down to Earth. It can then vaporize once again, rise, then crystallize and fall once again in a continuous cycle. Once this water vapor comes in contact with some form of particulate, it can then bond itself to the particle and transform into water. When enough form together, we begin to see this in the form of a cloud. Eventually, when saturation is attained, it falls in the form of precipitation.

Precipitation: Inputs

As the planet warms, the heat is built up in the form of energy. Storms are the most common and frequent ways the Earth releases this energy and aids in bringing about significant cooling to the surface of our planet.

I began to recall my time in the Marine Corps and spending time at the beach. The sand would be so hot, and the asphalt even hotter, and by the water, it was cool. Different objects have different variables of heat absorption, and in the sunlight, everything becomes warmer at different rates. Water absorbs and expels heat quickly and is seen every time a hot piece of metal is quenched in a bucket of water. Because water is our primary coolant that we use in such things as our cars, boilers, and chilling towers, I decided to see if precipitation levels oscillated along with our heating and cooling. If we were heating, then there should be an increase in precipitation, and the graph should correlate with past warming and cooling trends.

Minneapolis, Oklahoma City

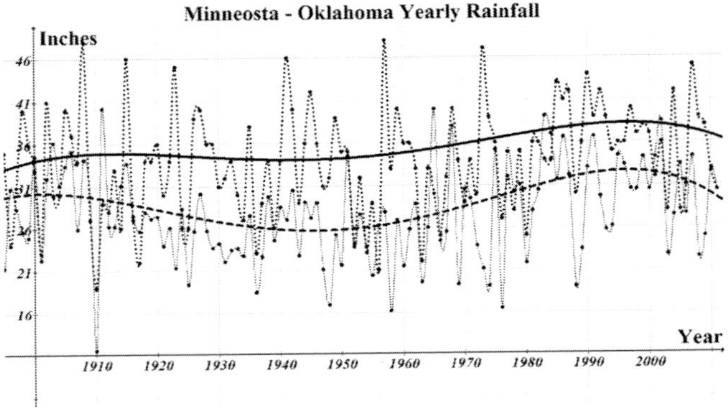

I decided to take two different readings from the Midwest so I went to Oklahoma City, Oklahoma, and drew calculations for their city as well. Subject to influences from the Gulf Stream and greater precipitation levels, they still demonstrated a decline in precipitation levels during the 1950s and 1960s, but not the extreme experienced further inland.

In 2012, North America experienced one of the worst droughts since the time of the great Dust Bowl. In Minneapolis, we had 29.59 inches in 2012. From 1971 until 2000, the overall yearly precipitation was 29.41 inches, indicating a normal season of precipitation. During the era of global dimming, temperatures and rainfall declined together due to less evaporation.

In *USA Today*, they reported the following: "For the year, the average precipitation total for the lower 48 (also known as the contiguous US) was 31.17 inches, which is 2.03 inches above the twentieth-century average. This made it the twenty-first-wettest year on record and the wettest since 2009."[3]

One inch of precipitation per square foot is .623 gallons. One square mile is 27,878,400 square feet meaning one inch of rainfall over one square mile is over 17.368 million gallons of water. The average rainfall for the year 2013 was 541,368,134 gallons of water per square mile. Multiplied by the 3,119,884 square miles of area, we find that 1.689 quadrillion gallons of water in the form of precipitation fell on the lower 48 states in 2013. Lake Superior is estimated to hold an estimated three quadrillion gallons of water.[4]

Minneapolis and St. Paul together make up 114.58 square miles and one inch of rainfall in the Twin Cities is just over 1.99 billion gallons of water. Lake Calhoun, a popular spot in Minneapolis, is 3.12 square miles and is estimated to hold 4.835 billion gallons of water within it.[5] This means that a rainfall of 2.43 inches over Minneapolis and St. Paul alone would exhaust all the water within Lake Calhoun, or fill it. Due to the size of other cities such as Chicago that is 234 square miles, a rainfall of just over one inch would be equal to Lake Calhoun, arriving in Chicago airborne. This water vapor in the air accumulates to levels equaling that of a massive airborne lake traveling across the land, often picking up even more moisture along its path.

This gives us a visual perspective of our planet's precipitation along with overall average totals. During the pollution era, the solar radiance upon the land and seas lowered enough to begin cooling, resulting in decreased evaporation and is demonstrated by the decline in precipitation levels during this era.

Storage

I decided I wanted to capture rainwater from my roof and use it through the hotter periods of rainless weeks that were beginning to develop. I needed to find out how many 55-gallon drums I would need to capture this. We cannot use the roof size because it is slanted and we can only use the area of land cover. The corner of the roof area came to 20 feet by 30 feet, or 600 square feet. At .623 gallons per inch, I would need seven 55-gallon drums to capture the 373.8 gallons of water for only one inch of rainfall. This was far more water than I had anticipated, so I began to do calculations within a size that my mind could more easily comprehend, my home. Calculating a yard the size of 105 feet by 105 feet, one-fourth acre, I began to run the calculations. This is a total of 11,025 square feet and at .623 gallons per inch there is 6,869 gallons per inch of precipitation on my entire yard. At an average of 30 inches of precipitation per year, we find that there are 206,070 gallons of precipitation that falls on my yard annually. Most of this water goes down the street, into a gutter, and then makes its way to the local Mississippi River.

In order to understand the full impact of water, we will take an average home and do some calculations. We will use a home that has 4,000 square feet of altered landscape encompassing the home, garage, driveway, shed, sidewalk, etc. A 50-foot-by-60-foot home and garage would be 3,000 square feet so this is a conservative size for a home today. This means that there is 2,492 gallons per one inch of rainfall, and the average rainfall per year in Minnesota is about 30 inches per season. This comes to 74,760 gallons of water that runs off the roof, driveways, and walkways down the street and out to the river where it heads to the ocean. I then looked around to all the homes in the neighborhood, in the world. My question was simple, what did the planet do with all this water before man channeled it all to the sea? How did it use it, store it, and transmit it? I looked at the two large trees in my yard and began to wonder, could the trees contain this much water?

Using the logging industry statistics, I decided to make estimates on these two trees. The primary logging season occurs when the soils become dry enough to get the trucks through, but not so dry that fires could result. Because the weights available will be for wood containing less water than its full capacity, adjustments will need to be made for the tree's ability to reach saturation capacity. First, we need to calculate the weight of the tree's dry weight.

Tree Calculations

The growth rate of a red maple is rated at 4.5 years per one inch in diameter.[6] By multiplying this rate by the diameter of this red maple, we can roughly estimate the age of the tree. By taking the measurement of the tree at breast height, we find this maple tree is 118 inches or 37 inches in diameter. This places this tree at approximately 166.5 years old, estimated germination date, 1848.

The elm tree is based on a growth factor of 4.0 with a 129-inch radius and a 41-inch diameter, making it 164 years old, a germination date of 1850.

These are very rough calculations without much competition available. Located close to the Mississippi, this would have been one of the early locations for logging and development that also corresponds to known history in this area. These would have been the new saplings left behind and grew after the logging came through and cut down the forests. In those days, clear cutting was the normal operation, allowing easy access and transportation of the lumber. Today, these two trees would be a good representation of what a typical Elm and Maple tree would have been like in the North American forests.

If we take into consideration that the forestry plants 1,000 trees per acre, can harvest upwards of 850 trees per acre, and a typical acre of forest contains 450 trees per acre, we can begin to draw some conclusions.

Red Maple Tree

The maple tree has a radius of 118 inches at breast height, making it 37 inches in diameter. Logging industry estimates this tree at just over eight tons at the mill. For firewood, they estimate this to be six cords of wood at 4,684 lb per cord,[7] making it an estimated 14 tons. The difference between these two weights—eight tons and fourteen tons—is that the mill excludes much of the wood that is usable as firewood. The primary part of a tree used at the mill accounts for only half of the tree, the other half would be found in the branches and the canopy, and much of this is not used in the firewood calculations either. This would increase its weight to an estimated sixteen tons above ground. The weight of the roots would also need to be included. Roots do not go down very deep as many people think and only reach down 10 to 12 feet. The roots however will span out anywhere from two to three times the canopy, depending upon the type of tree and conditions available. Because this is also the major concentration of moisture, it will also weigh the most by volume. We will estimate the entire root system at half the weight of the trunk, or four tons, one-fourth of the tree to remain conservative. So we can estimate this tree to be 20 tons unsaturated.

When dried, there is 1,004 lb lost per cord, 6,024 lb of water. This means that 21% of its total weight at the time of cutting is water. Saturation in the spring can easily bring the water content up to 50%[8] of its total weight, especially when sap flow is at its peak.

Water is 8.34 lb per gallon so we need to first calculate the tree's dry weight. At 20 tons at harvest and 21% moisture, we can determine that the tree itself is an estimated 31,600 lb with 8,400 lb of water (1,007 gallons). If this tree became fully saturated with water, there would be 15,800 lb of water or 1,894 gallons. This is an added storage capacity of 887 gallons that can be absorbed and released into the atmosphere by this one tree.

Elm Tree

The elm tree measured 129-inch radius, making it 41 inches in diameter, reaching upwards of 8.25 tons at the mill. It's estimated at eight cords of firewood at 4,456 lb per cord, or 17.8 tons. Using the same estimates as above for the canopy and roots we can estimate the overall weight of this tree at 20.63 tons or 41,260 lb.

One cord is 4,456 lb and 2,872 lb after it is dried. This means that 35% of the total weight at harvest is water. At 20.63 tons, the total weight of the tree is 13.4 tons and water makes up 7.2 tons, or 1,727 gallons.

At 50% saturation and 20.63 tons, the water capacity would be 10.3 tons or 2,470 gallons. This is a difference of an additional 743 gallons of storage capacity.

Calculating Changes in Water Tables

Differing statistics for forests are available today, but most use a rate of 450 trees per acre, a spacing of just under 10 feet per tree. The forests early in our history were much larger and far denser than the forests we see today, and these two trees would be the typical trees we would find in forests untouched by man. In 1927, scientists of the United States Forest Service Allegheny Experiment Station began recording data that has continued to this day in the Allegheny National Forest."[9] Larger trees need more area, resulting in fewer trees per acre. So we will use these trees as an example but reduce this number to 350 trees per acre.

These two trees average no less than 815 gallons of water storage capacity available in the early spring when the soils become saturated from the spring thaw. Not all 350 trees in an acre are adult trees so we need to reduce this. By decreasing this to 413 gallons per tree, a figure I had previously used to offset for the smaller trees—we will have a more realistic overall understanding of an acre of forest. Using these numbers, we can calculate that the total amount of water storage capacity when saturated by the trees in the forests is 144,500 gallons per acre. Considering that there is 27,138 gallons of water in one inch of rain over an acre of land, we can conclude that the forests are able to store up to 5.3 inches of water in the spring and this aids in preventing water runoff.

Large forests are quite different than much of the forests we see today. As the Allegheny forest research indicates, it requires 70 to 80 years for an upper canopy to develop and a lower canopy to establish itself. This would result in a much greater volume of storage available with the other vegetation such as ferns and bushes that would develop. Additionally, the thatch acts as a sponge, maintaining much of the moisture in the soil below. All of these figures up to this point are very conservative.

A 109-feet-by-109-feet lot, one-quarter acre, would need nine trees to be considered at a minimum 10% remaining natural land cover, and today, this would be considered a norm in the urban world. With 350 trees per acre in its natural state, there would have been 87 trees on a quarter-acre lot. There is no farm or urban development that even comes close to this level of vegetation. Growing up, we had many places to go as little explorers investigating the wilderness, but one by one, they vanished. Swamps and trails were soon replaced with parking lots and buildings.

Brush, shrubs, and grass were not calculated into the forestry figures, so none of this can be added in. The grass we maintain in our yards is known as green concrete by many in the fields of natural resources. Their short roots offer very little protection from the incoming solar radiance, evaporates water rapidly, and bares no resemblance to the deep-rooted prairie grasses.

In order to maintain proper water levels in a typical urban environment where 10% of nature still resides, the water capacity would need to be over 130,000 gallons per acre, or 32,500 gallons per average home. Cropland would be different and would need the full amount of over 144,500 gallons per acre. Additionally, the water would need to be filtered and released, carbon dioxide sequestered and converted to oxygen while providing shade and food for the area residents to do what the trees do naturally.

Much of this water becomes runoff and reduces surface infiltration resulting in flooding. The surface then rapidly dries, causing rapid evaporation of surface water, resulting in a decline of water into the subterranean environments. The water is not allowed the time to penetrate the soil the way nature's thatch in the forests would.

We make sure to clean our driveways, mow our lawns, and remove the leaves and thatch from the soil. This thatch maintains soil moisture, preventing rapid evaporation and adding a measurable insulating value to the Earth's surface by maintaining a cooler environment. It also provides protection from incoming rains by deflecting and preventing impact splash. It absorbs moisture and provides nutrients for the forest. Within the forest and prairies, there are insects and animals that burrow through the ground that aid in this infiltration process. Farms and urban development concentrates on eliminating much of this life, preventing this from occurring. Instead, we go to extreme efforts to make our lawns look nice and initiate laws to make sure everyone complies. We then go to extreme measures to ensure nothing digs or burrows through our yards.

Although the forests and prairies differ greatly on the surface, their rooting structures are just as deep and vast as the forests, and both require several decades to grow and mature. The long prairies located on the eastern edge were known to be so tall that only the hat on a man could be seen while riding a horse. Prone to wildfires, the fires would eliminate any chance of trees surviving for any length of time, resulting in a separation between the forests and prairies. The western edge of the prairies was short prairies and subject to drought conditions, resulting from a lack of precipitation from being on the leeward side of the Rocky Mountains. The prairies allow for vast infiltration and absorption through their rooting system, but do not have the resistance to the heat the way the forests do, resulting in faster evaporation rates than the forests. As a result, there is more evaporation in the prairies in comparison to the forests, and the average temperatures are greater during the summer months. Altering the water storage is only the beginning of understanding how man has altered nature

Transpiration

Just as we perspire, vegetation goes through a process called transpiration. Large trees can release up to 100 gallons per evening under ideal conditions. This process aids in the purification of our water, removing contaminants and providing moisture to areas that the atmosphere would take downstream by the winds. Altering this will also alter water tables atmospherically down stream from these locations.

Using the above figures, we can estimate what a forest would release on a typical one-quarter-acre home.

As the rains infiltrate the plains and forests, there is a process of evaporation and precipitation that repeats itself across the land until it eventually finds itself at the Atlantic. With the exception of Central America, where the land bridge between the Pacific and the Atlantic is narrow, all water moves from the Pacific to the Atlantic across the land. This is also evident by the ocean's tides, Earth's rotation, and the gravitational forces that move these winds from west to east when the oceanic jet streams are removed from having any influence.

One acre of storage capacity based upon previous calculations comes to 144,500 gallons, or 5.3 inches of water per acre. The thatch acts as a sponge, allowing the forests to saturate themselves over time during the spring thaw in preparation for its growing season.

During the spring, we will average 12 inches of rain over one acre of land. This comes to 325,668 gallons. We will reduce this by 1 inch to account for some evaporation and runoff. This leaves us a total amount of 298,529 gallons and an additional 144,500 of excess storage capacity, and leaves us with 470,168 gallons per acre available for transpiration under normal conditions and seasons. Divide this by the 90 days in the spring when we receive this precipitation, we can conclude that an overall average of 5,334 gallons per evening is released by an acre of forest. Divided by the 350 trees, we come to an average output of 15.24 gallons per tree on average. Because we decreased the water figures by nearly one half, we need to double this for an assessment of the large trees. These large trees release over 30 gallons per day on the average through transpiration, but can release up to 100 gallons per tree if conditions are ideal. The forests will only release excessive available water but will not reduce this amount beyond its own needs as water availability declines.

Based upon the average rate of 5,334 gallons of water being released per evening, per acre, and with a maximum capacity of storage at 144,500 gallons, we can conclude that the forest will exhaust all of its available water within 27 days without rain.

Because transpiration will be higher when water levels are at their peak and decline as they reduce, the forests will exhaust this rapidly and then reduce their rates until replenished with more moisture. Within a matter of a few short weeks, signs begin to surface as the vegetation becomes strained. Droughts are not a reduction in yearly rainfall amounts, although they do influence conditions, but are generated by untimely rains. After several seasons, a severe drought condition will begin to alter the landscape and climates affected, which is what occurred during the 1930s.

With these variables, we can determine that the forests can transpire vast amounts of water into the atmosphere when water levels are high, but can only sustain this for a few days before levels will begin to decline. When the rains come, they will increase once again in a continuous daily cycle that begins in the spring and ends in the fall. They offer ideal wind blocks and aid in moderating surface temperatures by reducing the solar radiance upon the surface. In these daily cycles, we find the air and water filtered, carbon dioxide sequestered, oxygen produced, water infiltrating into the sublayers, and massive amounts of wildlife present. They also provide protection for many species and offer food.

Transpiration allows for a slow methodical release of stored water in the form of vapor during the evenings and minimizes daytime evaporation. This, in turn, would have naturally produced more evening and early morning showers. This increase in water vapor would also maintain a level of warmth through the night, but the moisture released in the evenings and early mornings would supply the surface with cool purified water that would offset this rise in temperatures. As the night turns to day, the vegetation shuts down the release of water and prepares to begin a new day. The normal days would give way to clearing skies to maximize the solar energy necessary to sequester vast amounts of carbon dioxide through photosynthesis. This is a very harmonious relationship of our planet. The shade they provide for the surface would have also maintained a cooler temperature throughout the day. There would be vast filtering of air and water along with enormous amounts of oxygen being released, promoting growth and health all over the world and benefiting all life, including man.

Storage Deficiency

We have concluded that there is an estimated 932,105 square miles of converted forests alone in the lower 48 states today. Calculating that 10% of nature remains within this land, we can determine the storage deficiency. At 144.500 gallons per acre, less 10% comes to 137,050 gallons shortage per acre. There are 640 acres in a square mile, so we can conclude that there is 87.7 million gallons per square mile shortage. With 932,105 square miles converted, we can determine that the amount of storage deficiency within the lots from the conversion of forest, within the trees alone, comes to an estimated 81.745 trillion gallons. This number can easily double when we add brush and swamps into the figures and potentially even triple if we add the loss of the prairies into these calculations. This indicates that we are exceeding 200 trillion gallons of water shortage across the entire United States' lower 48 states every spring. This results in excessive water runoff that exhausts the supply quickly, especially in the early spring.

Underground Water Extraction

Underground water has its limits based upon physics. Just beyond two miles beneath the surface, the temperatures rise to a level that boils water. Once it becomes a gas, it will begin to slowly rise toward the surface where it can condense and become water again carrying the heat to layers above.

In 2005, there was a total of 79.6 billion gallons of freshwater withdrawn from underground water sources every day. Of this, 53.5 billion were used for agriculture, and 29.1 billion were used for other sources such as city water.[10] This comes to over 29 trillion gallons a year reduction in underground water sources.

> Surface water historically has been the primary source for irrigation, although data show an increasing usage of groundwater since 1950. During 1950, 77 percent of all irrigation withdrawals were surface water, most of which was used in the western States. By 2005, surface-water withdrawals comprised only 59 percent of the total. Groundwater withdrawals for irrigation during 2005 were more than three times larger than during 1950.[11]

In 1972, I saw my first irrigation system in place and wondered how it worked. It was a funny-looking contraption placed in the middle of the field. Since this time, it has become a farming standard in many areas around the world, and today we are installing them rapidly to offset drought conditions. Additionally, there are communities that have no surface water available and as these areas build, the demand increases their withdrawals, and there will be a rise in surface alterations. With a vast reduction of infiltration due to water runoff, decreased storage capacity, increased solar intensity and surface evaporation, this amount is accumulative over time, especially knowing that infiltration and surface storage has declined. If we take the year 1974 as a starting date, we can conclude that over one quadrillion gallons has been removed from underground aquifers over a 40-year period.

Water Runoff and Infiltration

Water runoff is a natural process, but when land is altered for crops and grazing, the amount of runoff and lack of infiltration becomes far greater. All the land west of the Mississippi in the northern region relies on the Pacific for moisture while east of the Mississippi is influenced by the Gulf of Mexico. The further east and south, the greater the impacts are felt from the Gulf of Mexico. As distance increases away from the oceans, the more dependent the land becomes on the repetitive process of rainfall and evaporation to move water like pumping stations along its path. Declining moisture levels along this path will result in a reduction of needed moisture to atmospherically downstream areas reliant on this water. Like kinking a hose, anywhere it is cut off will affect everything downstream.

In the forest, there are many areas of low land swamps that are vital for the surrounding environments. This natural watershed provides slow infiltration of water into the sublayers and maintains a higher water level for the area's vegetation. Nature also provided a creature to oversee many of these areas and assist in holding water back so that it could infiltrate more—this being the beaver. Beaver pelts were in high demand in the early days, specifically for top hats like Abe Lincoln's, causing their population to be wiped out. Some of these areas dry up quickly in the summer, while others can remain throughout the year. From a farming perspective and looking at available land size, it is far more efficient to create deep water lakes to store this water to maximize land area for agricultural use and get rid of the beaver, and this is what we did. This has resulted in a cumulative effect over time, causing a massive decline in subterranean water levels.

Runoff varies based upon soil type, but is greater when rainfall increases and the soil becomes saturated. Additionally, cropland is exposed to the elements and as a result experiences impact splash that results from rainfall. This tends to seal the surface, preventing infiltration that increases this runoff. The farmers, to decrease this effect, plow their field early in the spring to allow the water to infiltrate the soil. This runoff has been experienced in the form of flooding throughout America. This water that was once used by the Earth to carry away excess heat built up is now channeled directly to the sea. Throughout history, we have had to continually increase the heights of various levies across the nation to support this runoff effect.

In altering land for farm use, these lowland areas are filled and the water is channeled away to creeks and rivers so that a field's production can be optimized. The same applies to cities, and by doing so, we reduce the yearly flow into the subterranean aquifers adding to a continual reduction in the water tables.

Today, there are 907 million acres of farmland in America, 382 million acres for crop production, and 525 million acres for grazing.[12] So we can calculate that every inch of water runoff would equal 565 billion gallons.

Most of the surface water moves along the surface and only a small percentage of this water infiltrates into the deeper subsoil and aquifers. This results in a cumulative effect increasing over time. Using the figure of one inch loss per year over several decades results in a loss of 56.5 trillion gallons over a period of 100 years.

America's lower 48 is 3,119.884 square miles and 2,959,064 of this is land. If we consider that America's lower 48 is now 70% altered from its original state, this will leave us 2,071,344 square miles altered. One inch of water is 0.623 gallons per square foot, and there are 27,878,400 square feet in a square mile. This means that one inch of water over one square mile comes to a total of 17.368 million gallons. At 2,071,344 square miles and 17.368 million gallons per square mile, we see that the estimated loss of infiltration would be nearly 40 trillion gallons per year accumulative.

Totals

There has been over one quadrillion gallons removed from underground aquifers over the past 40 years (1974–Today) and this is rising rapidly with recent drought conditions.

Lake Michigan has 1.3 quadrillion gallons of water. Our answer to current drought conditions is to put in place more wells and extract more water to irrigate the crops. This only accelerates our problems today and our long-term effect will become catastrophic.

At just 1 inch per year, there is an estimated 40-trillion-gallon decline in water infiltration into the subterranean deep level aquifers per year or 1.6 quadrillion gallons the past 40 years due to runoff.

This is a 2.9 quadrillion gallons of depleted water over the last 40 years and continues to increase yearly, and this excludes previous years before 1974. This would mean that we have channeled away and pumped more water out of the subterranean reservoirs than the total amount of water found in Lake Superior.

Additionally, we need to include that every spring we begin the season with an immediate decline of an estimated 250 trillion gallons of water storage capacity across the entire lower 48 states as a result of the depleted forests and prairies.

We know that the Great Lakes create their own weather systems in the north, and the same would apply to the weather systems of the original forests and prairies having a direct impact on our weather patterns. Removing them has had a direct impact upon our surface weather, water levels, carbon dioxide concentrations and temperatures, but how much and to what extreme still needs further research.

Erosion (Declining Soil Nutrients)

Erosion and its impact will vary depending upon its natural environment, type of conversion, and the extent of those changes. Nature offers vast infiltration of precipitation by channeling water through its rooting system that penetrates the crust and the surface; thatch aids in maintaining moisture and preventing runoff. Insects and wildlife burrow into the ground also aiding in this infiltration process. Soil nutrients are replenished year after year, and when altered, these nutrients slowly become depleted from the soil.

Nature provides canopies that aid in slowing down rainfall and thatch that protects the surface soils and prevents impact splash. The thatch also maintains much of this moisture like a sponge, preventing runoff and preventing topsoil degradation.

In contrast, years of farming practices lead to compressed soils and a loss of nutritional contents that nature provides through their organic matter. The results of these practices are poor soil structure and increased runoff compounding this effect. Wind has an additional effect upon the soil blowing away necessary surface nutrients during dry spells where the forest and prairies offer protection. Nature's vast rooting system helps hold the topsoil and nutrients in place in comparison to farmland and urban developments. Landslides and flooding are often the side effects of overdeveloped land.

Heavy rains and storms produce additional runoff, washing away much of the necessary nutrients into the creeks and rivers, producing heavy sediments. Over time, this results in lower crop yields and increased heating effect due to the loss of water availability. Rotating crops can help aid in the nutrition compounds of the soil and offers only a short-term effect on the crop yields, but long-term degradation will continue. Water runoff is a natural process, but when land is altered for crops and grazing, the amount of runoff and lack of infiltration becomes far greater and is unavoidable.

Sand

Across the world in the midlatitudes in the northern hemisphere, we find the desert belt. This includes the Sahara, Gobi, and Death Valley to name a few. Sand is one of the planet's vital resources that absorbs heat rapidly, and also expels it quickly. When we spend time in the desert or on the beach, we can feel the heat radiating from the sand during the middle of the day, but at night, the temperatures drop quickly and significantly. Because of the principles of sand and its ability to absorb heat, it also expels it rapidly. The planet uses this as a blanket when the land becomes overheated and scorched over time. The sand protects the hard-packed soil and rock beneath it, insulating it from the daytime's scorching sun. The air between the particles of sand is an excellent insulator, and we use it in a variety of ways including storm windows.

As the heat rises in these regions, so does the wind and airborne dust. This dust becomes a vital part of the planet's cooling by distributing an increased amount into the atmosphere, allowing water vapor that rises due to evaporation to condense around these particles. As the heat rises, so does the volume of dust, wind, and evaporation proportionately. These are the three necessary ingredients to create clouds, offer shade, and release the rains to cool the planet's surface.

Summary and Continental Impact

Early last spring, I was on a service call on the Canadian border. While I was there, I noticed a scent in the air—a scent I had smelled before, but could not place where or when this was. As I returned, the scent soon vanished and I remembered when and where I used to smell this. It was up north at my uncle's resort in Cass Lake, Minnesota. It brought back many wonderful memories. The scent was coming from the forests and their production of oxygen in the early spring, a scent that we no longer have in our air. It's a scent that when you smell it, you want to keep inhaling it. It's so very refreshing.

America's lower 48 states total 3,119.884 square miles and 2,959,064 of this is land. If we consider that, America's lower 48 is now 70% altered from its original state. This leaves 2,071,344 square miles altered from its natural state. Considering half of these miles were previously forests, we can determine that 1,035,672 square miles or 662,830,080 acres of land were originally forest and this was used in this research for land calculations. Using the calculation of 10% remaining in our cities, yards, and areas on the farm, we can determine that over 596 million acres, 932,105 square miles have been uprooted since America was colonized.

The US Forest Service today declares 751.2 million acres of "forest land" remaining, which indicates the numbers being used in these calculations are very conservative. It appears there is far more forest land than prairies. What is needed to qualify as forest land is at least one acre of land with 10% occupied by trees of any size or "formerly having had such tree cover and not currently developed for non forest use."[13] Some examples are land not permanently altered for the use in crops, improved pastures and even some residential areas.

Using figures from the EPA on agricultural use, we find that today there is 907 million acres of farmland in America, 382 million acres for crop production, and 525 million acres for grazing.[14]

Forests offer great protection from the wind and create a more stable ground environment. On cooler days, they offer warmth while maintaining a cooler environment on warmer days. In this way, they act as a buffer, maintaining a more stable environment. During the summer, they cool the planet's surface by 10%–30% depending upon its vegetation and location by providing shade from the incoming solar radiation and maintaining a stable and moist environment.

By filtering the air and water through its natural processes, they are able to trap many elements within its cell membranes as the water flows through the tree, purifying the water in this process. Bacteria and other elements can become trapped within the tree, thus aiding in the cleansing of the Earth's water supply by encompassing this within its own structure.

There is a decline in water levels due to increased runoff and erosion. There is also a decline in water tables due to extraction with no positive numbers available. Mathematically speaking, we are drying up in a very slow methodical way.

Every year, we cut more and more of the Earth and try to form it to our imagination, yet today we can't even imagine the pristine beauty that was once available. As a result, we are drying out the surface soils by increasing runoff and eroding the soils.

A car or nuclear reactor overheating may run a little further or longer in the dead of winter when it's extremely cold, but without coolant to provide circulation, there is no amount of external force that will alter its outcome. If coolant levels decline below the needed levels and create a shortage in circulation, heat will rise resulting in an increased surface temperature.

Man's response to our drought conditions is to drill more wells and to bring even more underground water to the surface, which in turn will only accelerate this problem.

The water cycle is an essential part of the planet's cooling, and when restricted or reduced below its needed levels, a greater potential for breakdown occurs. When we are talking about heat and pressure, there are also stress factors that need to be considered. The Earth is a sealed system, like a combustion chamber, and increased heat and pressure can stress even the hardest of metals to cracking and even meltdown.

It's in this knowledge that I decided to shift my focus toward earthquakes and heat gradients of the earth. I had run some statistics several years ago that demonstrated the correlation between oil production and earthquakes. If there is another source causing a rise in earthquakes, then this needs to be considered in the research. Because we are confirming the planet's expected response to known physical alterations, then they need to be considered. This research was formulated in 2009 prior to the North Dakota oil rush.

EARTHQUAKES AND PASCAL'S THEORY

Although the US Geological Survey (USGS) indicates that there is no correlation between earthquakes and the exploration and extraction of oil, the physics involved seem to oppose this position. Oil and gas are the result of organic material under intense heat and pressure over a period of time. Oil's abundance made it an ideal candidate early in our history for use. As time has passed and the demand continues to rise, these resources have been slowly depleted. It's this cumulative effect on the land that was researched, and I will let the reader and this research speak for itself.

All the figures used below were retrieved from the USGS and the Energy Information Association (EIA), both government agencies responsible for documenting earthquakes and oil production.

Even as a young boy, I often thought about what would happen when you suck that water out from underneath us. Certainly, the ground would have to go down due to displacement, but it should fill back up too, shouldn't it?

Later in life, I learned about the principles of hydraulics and Pascal's law. In general, his law states that a liquid in a sealed container cannot be compressed and that pressure is exerted equally in all directions in this environment. This is how brakes and automatic transmissions work in cars and other vehicles. I could see how we would create problems for ourselves eventually if we remove enough water from the sublayers. Like walking next to the water by the beach compared to walking in the sand dunes, same soil, just absent of its moisture. As mines have demonstrated throughout history, the Earth wants to fill in voids beneath the surface due to the extreme pressures experienced as you travel deeper. But then I looked at the oil industry and saw something even more concerning. Could oil extraction, exploration, and forced injection used to maximize an oil field's production cause earthquakes?

Oil production could have an influence on large-scale earthquakes over time, but are known to periodically produce small-scale tremors. Earthquakes work proportionately, meaning that a number of small earthquakes need to occur before a larger earthquake takes place. For example, there may be 100 level 3 earthquakes before one level 4, and 100 level 4 earthquakes to occur before one level 5 and so on. If oil production does indeed influence small-scale earthquakes, then mathematically, the probability of a large-scale earthquake must rise proportionately with it.

As we go back in time, there is more speculation, but both the monitoring and technology improves every year. I decided to use 1980 as a starting date, a time when every earthquake within these variables should have been detected. I took all the earthquakes in the world between the levels of 3.0 to 5.9 and graphed them out.

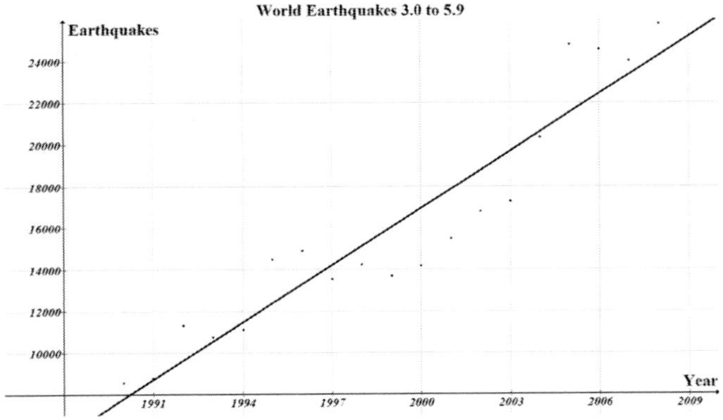

We seemed to have come to a time where we are now experiencing more earthquakes than ever before. There is credence that monitoring and population densities have increased awareness, but this does not account for the past 30 years. Many of the oil beds around the world have been consumed, and these are areas that would be considered safe to drill for oil because they are not along any fault lines. Today, with the reduction in these fields and greater demand, they are now expanding exploration in and around these tectonic plates, particularly of concern, the Caribbean and the Far East.

Since 1990, we can see a sharp increase in only 20 years—from less than 9,000 in the early 1990s up to nearly 25,000 today. Although it is known that drilling and extracting oil often produces earthquakes in the magnitude of 3 to 5, it has been noted that there is no significant rise in larger earthquakes around the world.

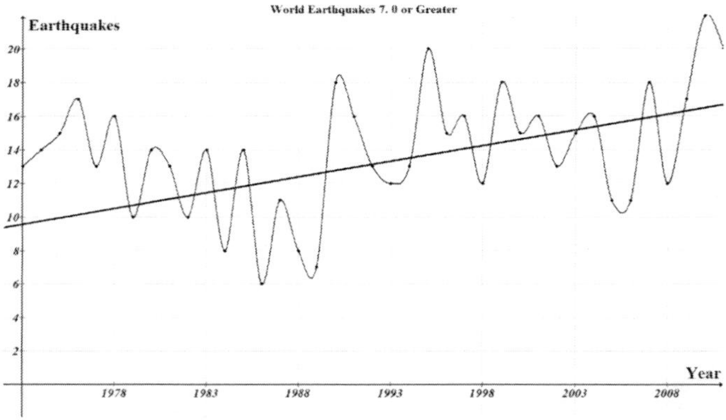

Earthquakes 6.0 and greater were less than 110 per year back in the early 80s and are now exceeding 160 a year. In this graph of earthquakes above 7.0, we can see that there were ten recorded earthquakes exceeding 7.0 or above in 1980 and now averaging over sixteen annually. In 2010, there were twenty-two.

Although many continue to speculate other causes or influences on today's earthquakes, I decided to look at

Alaska and graph this area out. This was virgin soil prior to 1957 and could be used as a good base model.

Logic tells me that extracting one barrel of oil from the ground will have a minimal effect, but extracting 100 barrels would have 100 times the effect. The effect is directly proportionate to the area of mass that lay above and around the oil and gas being extracted. Extracting these elements will cause pressure changes below the soil, causing a displacement. When extracted in or along fault lines, where a great amount of oil is found, there should be an anticipated increased effect resulting in earthquakes near these areas. Eventually, at some point, a precipice will be met when the amount will eventually have an effect that will destabilize the plates. If this is true, then we would expect to see production increasing faster than earthquakes, and where drilling subsides, we would expect to see earthquakes decrease slower than the production decreases, a delayed event.

Alaska Earthquakes And Oil Production 1981–2009
1981 Beginning of Production

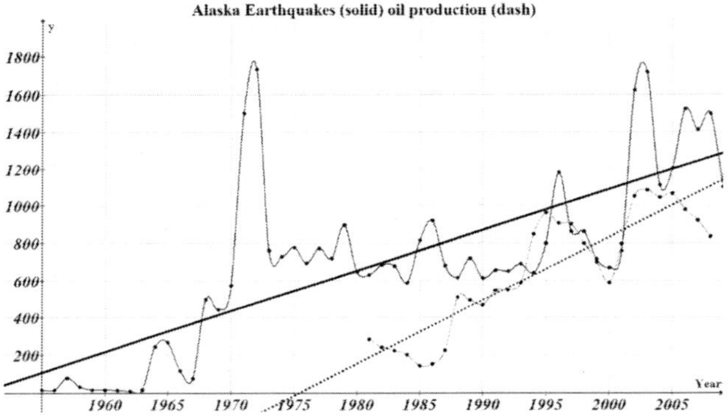

Dashed line indicates oil production, solid line indicates earthquakes above 3.0

In 1964, the Great Alaska earthquake took place. Afterwards monitoring for earthquakes changed and is demonstrated by the graph's numbers for that year. After 1972, the land became more fragile and susceptible to earthquakes, and there also seems to be some clear indicators that looking for oil also causes earthquakes, specifically 1971 and 1972 before the pipeline began. There is also a bump in the graph in 1957 when oil production first began and in the 1960s when they expanded exploration and extraction. As this graph indicates, increased earthquake activity occurred after the pipeline began and responded directly to production—specifically, the rise, fall, and rise again in both production and earthquakes from 1996 to 2002.

Oil production around the world has been on the rise except for one area, the lower 48 states. Oil production had been declining in the lower 48 year by year, and if this hypothesis is correct, then we should see a decline in earthquakes of 3.0 and above for the lower 48. Many oil wells were capped off and production drastically slowed up in the early 1990s due to the low cost of oil. Here are the results with the dashed line showing the average production while the solid line is the average trend of earthquakes since 1990. The lower 48 is the one area of the world where we can find a reduction in both earthquakes and oil production.

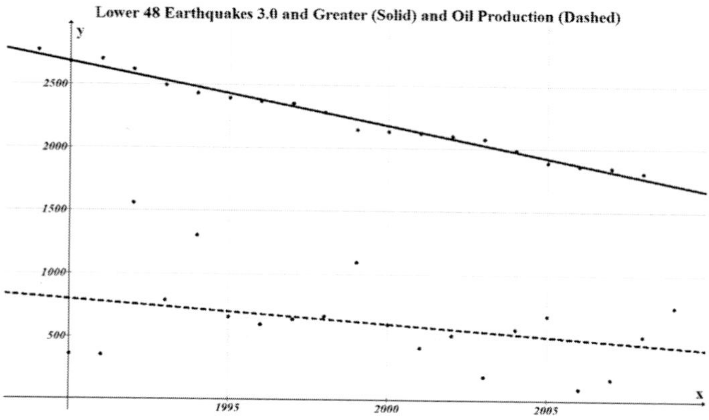

In this graph, we can clearly see the decline in earthquakes; the only large-scale area on the earth showing a reduction. This demonstration conforms to known physics and the results confirmed my expectations.

Since this time, there has been massive production coming from North Dakota that will be discussed later, but it is also clear that these levels of earthquakes have risen drastically in direct correlation to production. Further research will only confirm this. Understanding the physical properties at work, we can only expect more and greater earthquakes here in the lower 48. This is being accomplished in order to fuel the financial greed of a select few people who will ignore any responsibility or liability for their actions.

HEAT GRADIENTS AND OCEANIC CIRCULATION

Our planet's primary cooler is water, and the greatest quantity is found in the deep oceans. As the axis of the Earth alters for a change of seasons, one pole freezes while the other thaws. This is a vital cycle, allowing for the fresh cold water to sink to the bottom of the ocean's salty waters constantly replenishing the cool water supply deep into the oceans. This gradually warms and rises to the surface, producing our surface weather and jet steams known as the Great Ocean Conveyor. It's this difference in the ocean's temperatures between the poles and the tropics that drives this conveyor, and both the daily and seasonal changes continually refuel it. This alters and moves the surface jet streams and is the major driving force in our weather and the primary source for all of our precipitation.

This cooling is evident by the deep ocean temperatures averaging 34 to 39 degrees and makes up 90% of the oceans. The Mariana Trench is the deepest part of the ocean over 36,000 feet deep and the temperatures range from 34 to

39 degrees. In contrast, inland, we find a stable layer just beneath the surface averaging more than 10 degrees above these temperatures and incrementing one degree every 70 feet in depth. This means that inland, at 1,000 feet above sea level, if we were to drill down 37,000 feet, we would be at the same level as the Mariana Trench. The temperature we would find here would be over 583 degrees in comparison to 39 degrees in the ocean. These temperature deviations demonstrate the cooling effects of the oceans, much the same way we use a cooling tower for a nuclear reactor.

This produces heat and pressure under the continental plates that forces them to rise above the cooler ocean plates. This heating effect and temperature gradients allow the ocean plates to ride beneath the continental plates, creating what is referred to as subduction zones. This heat and pressure from the core maintains the tectonic plates in a continual motion along the Earth's surface and is known as the continental drift.

Deep water oceans cover more than 60% of our planet. In these waters, we find temperatures are less than 40 degrees. If we add the areas with permafrost, we will find that nearly 80% of the earth remains below 40 degrees and the further north and south we travel, the deeper the impact becomes.

If we also include seasonal shifts, the greatest period of cooling occurs during the summer months in the southern hemisphere. This is due to the massive amount of land experiencing cooling during this time frame in the northern hemisphere. During this time, we are also closer to the sun that offsets these alterations, creating a balanced effect.

DISCOVER EARTH'S THERMAL SWITCH

I had a continuing education class that interrupted my research that I had to attend for my state licensing requirements as a low voltage contractor. I proceeded to put down my research and go off to school. I met several other technicians from various industries around the state and did my best to enjoy the day. One field was a group that worked in the heating and cooling industry. It was on this day that I finally understood how the Earth was warming, and it brought clarity to all of my research along with the proof and evidence supporting this theory that I was seeking.

We began to discuss Ohm's law, a formula I have used my entire life and then someone in the class made a statement indicating that heat had a mirror equation that I hadn't considered. My mind began to revert back to the heating field and it all came together in a flash. They both used resistance in their formulas to calculate energy flow, and they are both balanced equations. If there was ever a time when it appeared as if a light bulb went on in someone's mind, this was that moment. It brought everything together and now is not only understood, but measurable too.

In electricity, decreasing resistance increases the flow of current, and decreasing resistance in thermodynamics will result in an increased temperature over time. I was always taught to look at it like water and a faucet. The water is your flow and the faucet is your resistance. As you open the valve there is less resistance to the water pressure and this results in a greater amount of water. Additionally, the water goes from the faucet to the sink and not the other way around. Time is a variable when storage is needed, even with the water. It takes time to fill a jug of water or charge something, like your phone. It also takes time to heat your home, or in this case, the world. The answer was found in the basic laws of thermodynamics, which also encompasses all of Earth science. This is an equation providing clarity, reasoning, and understanding to everything I had found over the years. For three years, I had been researching various aspects of our planet, and at this moment, it brought about clarity and now I could see how it all fit so neatly together, like a puzzle waiting to be assembled.

In electricity, energy can only flow in one direction. Energy in the form of heat can *only* flow from hot to cold and cannot flow from cold to hot. All of science understands and agrees with this basic law of thermodynamics, but no one has ever asked the right question. Because this happens out of sight, it has been easily overlooked.

Now I know many of you do not understand the evidence supporting this theory yet, so let me explain it in the question I posed to my father. I called him and I knew if I could ask him the right question properly and fully, his answer would bring clarity to the planet's heating and cooling, and he would immediately understand what I have been searching for. I felt that if I could explain it in a way that he could understand, then I could explain it so that everyone can. When he answered the phone, I told him I had a question I wanted to ask him. He told me to go ahead and ask away, so I asked him this question:

"Dad, there's a stable thermal layer beneath our feet that science says remains approximately 50 to 55 degrees year around. The core gives off heat 24/7 below it. When the area above this thermal layer reaches a temperature above the temperature of the thermal layer below, let's say 70 degrees, what happens to the heat coming upwards from the core?"

Without even a pause, he replied, "It becomes boxed in and it waits for winter."

There was the answer! This is the key to unlocking all of the answers. The source of our increasing and decreasing heat cannot be stored or measured in our atmosphere, but a cycle that occurs every year beneath our feet. When excess heat is built up underground over time due to the alterations on the surface, climate alterations begin. This is just the beginning of understanding how the climate changes. The interesting part is how carbon dioxide and the atmosphere effects this cycle and will be discussed later. I then asked him if he now understood how the planet heats and cools. He chuckled and said, "Yes, but what am I going to do about it?" He was right, there is nothing any one person can do. It's something that must be addressed on a global scale.

I then turned to a group of technicians and I asked them this very same question, and one of them spoke up and said, "You're right!" He looked at me with a look of enlightenment and wonder and said, "It has to…because it's the law." He was right, it's one of our basic laws that the Earth gives to us and is irrefutable.

Because of the simplicity and the fact that this occurs in an area out of sight, it has been overlooked for so many years. We have mapped the Earth, the moon, and even the surface of Mars, but even our own deep oceans remain virtually uncharted. We tend to look to what we can see, but what we can't see often evades us. Because it is one of our planet's basic laws, and these laws govern Earth science, then Earth science must adhere to this law.

Now I knew there was one way I could confirm this. If I calculated the number of days when the temperatures remained above 50 degrees without ever dropping below this temperature and plotted them out on a graph. If the graph oscillated with the temperatures, it would confirm this yearly cycle. I then applied the same five-year timeline to see if this graph of "time" would confirm that the law of thermodynamics works within the Earth.

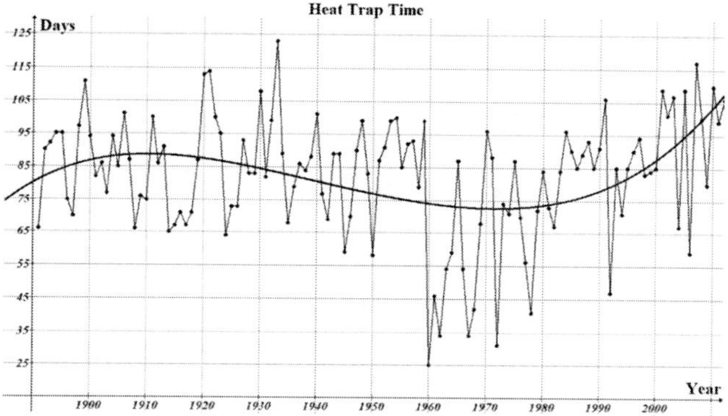

As I plotted out the graph and placed the trend line, this graph unfolded. It was a mirror image of the temperature variable indicating that the length of time the land boxed in the heat has a direct effect upon our planet's surface temperatures. Finding the source is only the first step in understanding global climate changes. This explains the gradual build and decline we see in the graphs over time. This also explains the data I had researched these past few years and the discoveries I had found along the way, supporting everything the reader has read to this point.

Heat from the core cannot penetrate the surface when the surface temperatures are greater than the layer below. This thermal layer is our planet's primary resistance protecting the surface elements from the heat rising from the core. If this were to breach our planet's surface, we would become the mirror image of Venus. The actions that prevent this from occurring are the planet's axial tilt altering our seasons and the speed of our rotation, something Venus is lacking.

When summer arrives, the surface acts as a thermal shut off valve, and the energy flow reverses. The energy coming from the core below begins to build and stores this energy in the form of heat with no outlet, except for water. Water can move this heat through underground basins, swamps, lakes, and rivers. Anywhere we find the temperatures lower than the sublayer below, this area will act as an exhaust vent, allowing it to release this heat. The amount of heat released will be directly proportional to the differentiation in the two temperatures and the size.

In the summer months when the heat comes upwards from the core and comes in contact with water, it travels immediately toward the surface. When it reaches the surface, there is no outlet because the surface temperatures are greater, so it remains boxed in. At night, all the vegetation on the surface opens up and allows the release of this subterranean heat in the process called transpiration. This process is not cooling the vegetation, but the Earth. Where temperatures remain above 50 degrees at night throughout the year, the vegetation remains intact and allows for continual venting.

By boxing in this heat through the summer months, it naturally stores this for the roots and life of the forests and prairies to sustain throughout the winter months. I thought, if this didn't exist, it would jeopardize all of life on a continent or even the entire planet. I saw this as a harmonic cycle of the planet that had been established since the beginning of time, something that mankind has altered greatly. Mankind has repeatedly caused this same effect before but on smaller regional scales. We do not directly warm the planet by our activities, but alter the way the planet responds to the sun.

This is part of the planet's normal cycles, and there is a purpose behind this. Our longest day is on June 21, but the land locks and maintains a low temperature of 50 degrees in Minnesota on an average of June 14 over the past 112 years. If this cycle was not present, we could potentially drop to less than 50 degrees by June 28, instead the average date is August 29. The planet uses this energy to support the life in the forests and prairies by extending the growing season. In the fall, the leaves drop and the trees seal themselves in preparation for winter, keeping in and maintaining much of this warmth for all life that goes underground to hibernate. The snow then acts as a blanket over the surface protecting it from subzero temperatures, preventing it from driving too deep into the ground. This becomes evident in Minnesota as water lines continually freeze in parking lots and streets due to their direct exposure to the extreme low temperatures.

If this did not occur, there would not be enough time to reproduce and grow, and they could be subject to a deep freeze that would otherwise wipe out entire regions. Without it, there would be far more lifeless areas around the world and a narrow ring around the midlatitude that would support the diversities of life we see today. This allows for large scale moderation, a form of a buffer, preventing rapid heating and cooling and allowing for large amounts of land surface to sustain life. Life on our planet has evolved over time to adapt to these cycles and is a necessary cycle to support the vegetation and life on the planet.

I began looking up the founder of this law and his work. The law that I use daily is Ohm's law, and with heat, it's called Fourier's law. If resistance with either element rises, energy flow reduces; if resistance declines, then energy flow rises. With thermodynamics, this means the less resistance between the sun and the surface means more heat, resulting in higher temperatures over time and after passing multiple cycles.

As I researched Fourier's law, named after the founder, Joseph Fourier (1768–1830), I found he was the first to identify the greenhouse effect in a paper he published in 1824. He had determined that the sun was unable to heat the Earth to the temperatures we experience and had drawn the conclusion that the atmosphere maintained some resistance to outgoing heat. He maintained that the core was leftover energy that was going to deplete and was leftover from its origin, but science today has a different position regarding this theory, and this research expands even further upon this. Both he and science today continue to disregard the core's impact upon our surface temperatures, overlooking this cycle and its impact on our temperatures and weather patterns.

He was also one of the leading scientists in thermodynamics to identify the characteristics of heat, including thermal conduction and heat flow. It was through his research that I discovered this cycle in his "Analytical Theory of Heat," published in 1822. It was in his work of understanding the dimensional analysis of heat that he was able to formulate mathematical equations for heat that we still use today. In this theory, if heat remains constant and resistance declines, temperature must rise, and if resistance increases, then temperatures must decline. The irony…189 years later.

Briefly summed up, there are three laws regarding heat that relate directly to the Earth's greenhouse effect.

> Newton's law of cooling is that when two objects of different temperatures come in contact with one another, ultimately they will come to a common temperature with the surrounding environment. This is known as thermal equilibrium.

> The first law of thermodynamics is that energy cannot be created nor destroyed and what is lost in one process must be gained in another. This is known as the conservation of energy. This includes the relationship of temperature, volume, pressure and heat in both a sealed and unsealed environments.
>
> The second law of thermodynamics is that heat flow is unidirectional from hot to cold and cannot flow the other direction. Additionally, whatever energy is transmitted from one, there must be an equal amount of loss in the other. This is known as the direction of conserved energy.

Heat always moves toward a cooler environment, and in nature, this is always at a higher altitude. Even the core of the Earth continuously emits heat that rises toward the surface where pressure and temperatures become less, and this is where "heat rises" originated—in the heating and cooling industry. It's the natural flow.

Newton's law indicates this sublayer will continue to rise in temperatures in an attempt to find thermal equilibrium. It will, according to this law, continue to rise in temperatures that would eventually exceed the surface temperature in order to breach through. Due to the Earth's axial shift causing a change in the seasons, it is never long enough for this to occur. As this energy comes from the core in the form of heat, it will also follow the laws of heat with in a sealed container including a rise in volume or expansion, temperature, and pressure.

Atmospheric resistance is anything that comes between the sun and the planet's surface. This would include cloud cover, pollution, volcanic discharge, or anything that would deplete the solar radiance upon the surface, but in the air. Ground resistance is anything that protects the raw soil from the incoming solar radiance such as forests, thatch, wetlands, or prairies.

Based upon thermodynamic laws, the following is a known response to incoming solar radiation:

- When ground resistance or atmospheric resistance rises, surface temperature will decline.
- When ground resistance or atmospheric resistance declines, surface temperatures will rise.

When one rises and the other declines, then the resistance offering the greatest influence will determine the outcome of the surface temperatures. They will rise, fall, or remain constant depending upon resistant variables. Here are examples we have experienced:

- If atmospheric resistance rises more than ground resistance declines, then the surface temperature will decline proportionately to the difference between the two resistances. (Air Pollution Era)

Pollution caused a reduction in solar radiance that was greater than the increased radiance due to the loss of ground foliage. This resulted in a decline in surface temperatures.

- If atmospheric resistance remains constant (clean air) and ground resistance declines, then the surface temperature rises. (Pre-1930s and post EPA era beginning in 1970) When atmospheric resistance remains clear and constant, but surface resistance declines, temperatures rise. This has been repeated many times in our history.
- If heat declines and resistance remains unchanged, then temperature declines and this is proven through two events, the Maunder and Dalton Minimums—two periods of time that, for unknown reasons, solar activity declined.

During the spring and fall when the surface temperatures oscillate between day and night, so does the energy in a trap and release cycle. During the winter, the surface temperature becomes low enough to allow a continuous release of this energy throughout the season and is a vital cooling component for the planet in order to balance the environment. The amount of energy that flows between two points is determined by the resistance in the soils and biomass that create resistance, and the differences in temperatures between the two along with the period of time. The cooler the surface and the longer it remains in a cooled state, the more energy is released and the cooler the temperatures become on the surface over time. Although surface temperatures can often reach well below zero, snow cover offers an insulating value, maintaining a constant temperature near the freezing point of water on or near the very surface. This maintains a constant rate and the period of time the heat is allowed to release has a far greater impact.

Heat from the core differs greatly than the heat from the sun. First notably is that heat coming from the core is transmitted through convection. Because heat rises, all heat comes from the ground when it becomes frost-free in the spring. On a sunny day, it is estimated that 1076 watts of energy is received per square meter using the rate of 100 watts per square foot. The core is estimated at only 0.065 watts per square meter and is shadowed by the sun, but efficiency and storage is uniquely different. Although this 0.065 watts is in question, we will use this figure to demonstrate the effects.

The sun has far greater intensity, but there are many additional factors. For example, the sun offers no heat at night, reducing the solar energy by 50% over the course of a year. The energy from the sun is only able to penetrate down a very short distance as a result of our daily rotation and resistance. On overcast days, clouds continue to reduce this energy upon the Earth's surface significantly. On the surface, there is natural vegetation and ground clutter in the forests and prairies that intercept this energy and use it for photosynthesis, and this is where we find vegetation battling for the sunlight. The vegetation and the Earth's surface have a symbiotic relationship, protecting the Earth's surface from the sun. What sunlight penetrates through the forests and prairies quickly finds resistance at the ground level from leaves and debris, preventing it from having direct contact with the soil. Temperature differences in this thin layer alone have been recorded at over 37 degrees when exposed to direct sunlight for a period of time. The core is not affected by any of these conditions and remains 100% efficient.

This past winter in Minnesota, we had many water lines freezing six feet below the surface. When temperature samples were taken, the water was running at 33 degrees at the tap early in the morning. In the spring, the same water samples were taken and the water was running 43 degrees, but the frost line was still in the ground at two feet. With a frost line at two feet, only subterranean heat could warm this water. The core had already warmed this water 10 degrees higher than the layer just a couple of feet above it. Eventually, they find equilibrium and then lock when the temperatures remain greater than the sublayer below, and early in the spring, this would be an estimated 50 degrees on the surface.

The seasons also alter this effect, so the further you travel toward the poles, the less effective the solar radiance is in comparison to the tropics. So although the core provides far less heat, it is also 100% efficient regardless of the surface environments, but when compounded with a sealing effect throughout the summer months, it becomes a large and vast storage supply for heat, resulting in a slow methodical rise in temperatures creating what is known as the greenhouse effect.

In its natural state, the Earth will build this energy beneath the soil during the summer. The heat will be greater over the area with least resistance to the incoming solar energy, and less where the insulating factor is greater. In this situation, the forest floor will absorb less solar energy into the ground, resulting in less heat being released. The prairies, with a thinner insulating cover to the ground surface, will allow greater thermal absorption resulting in higher surface temperatures. As a result, the prairies will trap the heat into the sublayer earlier than the forests due to their exposure and reduced insulating effect. When winter comes, the surface temperatures decline and allow this built-up energy to escape. In its natural form, this oscillates every year and had achieved balance that man has and continues to alter. [1]

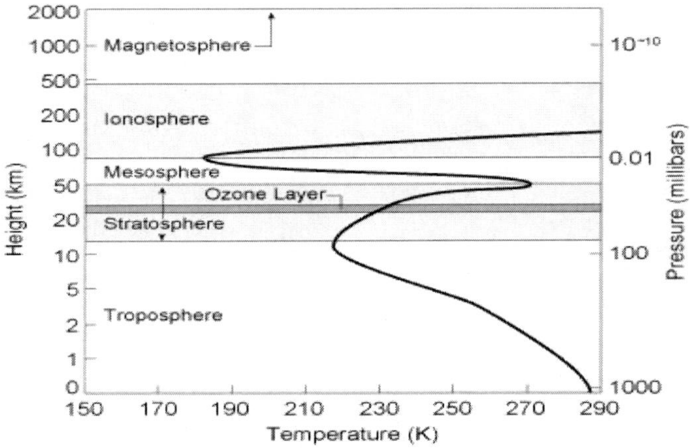

Between the sun and the surface of the Earth, there are three temperature gradients of heat, and each one indicates a resistance to incoming solar radiation. Based upon thermodynamic principles, anything offering resistance will demonstrate a rise in heat. The first layer of resistance is the magnetic field that blocks a vast amount of energy, and as a result, we see a rise in heat in this layer—the thermosphere. For the solar radiance that makes it through the first layer, it soon encounters the second layer—the ozone layer. This layer is responsible for filtering out even more harmful radiation from penetrating through down to ground level. The final layer solar energy encounters is the atmosphere where we find clouds and other elements that reduce the amount of energy even further before reaching the surface.

Temperature levels here on Earth are taken in the shade. If we were to take the temperature in the sun, the temperature would also reflect the radiance upon the mercury, raising the variables and adding to the common temperature of the surrounding environment. For the purpose of this research, monitoring the temperatures in the sunlight would have been beneficial over the years. This would have indicated a differentiation in resistance levels to the incoming solar energy.

On the Earth's surface, the incoming solar radiation, in comparison to space, has little effects on such things as a thermometer due to the resistance and filters the radiance encounters before it reaches the surface. As we rise in altitude and the resistance to the incoming energy declines, the more the mercury is affected by the incoming solar radiance.

In space, the temperature, without any filters reflecting and obstructing the solar radiation, causes items to rise substantially. If we wanted an accurate temperature reading in space, we would need to do the same thing we do here on Earth—find some shade and take some readings. We can easily find this on the dark side of the moon and find that the temperature in space is –250 degrees F.[2] On the surface of the moon in the sunlight, we find the temperatures rising to +250 degrees F. This demonstrates the amount of solar radiation being generated from the sun and its effects upon different objects depending upon their individual rates of absorption or reflection.

Man's Influence

When the surface is altered for man's use for farmland or urban development, the insulation is removed, exposing the surface directly to the solar energy and increasing atmospheric carbon dioxide concentration levels. In doing so, it accelerates the heating effect on the surface during the summer, trapping the heat quicker in the springtime, acting like preheating an oven. It also channels the water away from these areas, reducing the amount of ground water and infiltration that amplifies this heating effect. Over time, it then maintains it longer and drives the heat upwards toward the surface. Carbon dioxide levels accelerate in the atmosphere, causing the heated portion of the midlatitude atmosphere to swell, creating a feedback that also aids in accelerating this effect.

Over time, this leads to a delayed onset of the fall and declining winter seasons. If winter seasons decline, then there is a buildup of the heat below that will carry forward into the following spring when the heat becomes boxed in again. If the surface remains unchanged over a period of time, this heat rises in both duration and amount and can be registered on the surface through temperature changes. Over many years, the constant temperature will slowly rise toward the surface and the sublayers will increase in heat and pressure. This creates an earlier thaw due to the lack of time needed to dissipate the amount of heat stored, resulting in early spring floods and late season droughts. Over many years, this heat builds and continuously rises, creating its own feedback and is known as global warming.

There are many different numbers calculating the amount of heat that the core dissipates through the surface. One of the most recent and lowest assessments is 65mW per square meter of power being generated from the core through the land surface. For every day this heat is trapped below and the cooling season is reduced, there is a carryover effect that can be measured.

In the northern hemisphere, we find the massive amount of land and the greatest alterations that have taken place. We will calculate the natural storage and release for a given season. The amount of time is vastly increased as we travel south and declines greatly as we travel north, so we will use an overall average of three months (June–August). The core releases 0.065 watts per square meter, and there are 2,589,990 square meters to one square mile. This comes to 168,339 watts, or 168 kilowatts per square mile. There is 44,000,000 square miles of land in the northern hemisphere so we know that the planet stores 7.4 billion kilowatts. Watts are calculated in hours so the total storage would be multiplied by 24 hours to come up with 177.6 billion kilowatts per day of storage, or 177.6TW of energy stored below the surface every day. This comes to nearly 16PW of total energy that would be considered naturally stored during the summer months to maintain enough heat to extend the growing season and maintain warmth below the surface through the winter. As we can see, something that seemed so insignificant can now be understood as the power below.

If we assume that 16PW (16×10 to the fifteenth power) would be the natural average, and over time, this has expanded by 22 days we can determine the estimated amount of additional heat stored at 3.9PW. This would be carried forward and continues to build year by year.

Although these figures are speculative and based upon a source that may be in error, the heat formula and alterations are accurate and demonstrate how this energy rises and falls. No matter how much energy is coming from the core, the same results will occur, just at different values because of a balanced equation.

Because the surface is sealed and there is no wind to carry the heat away in the same way it does on the surface, subterranean water does. This would need to be added into any equation in determining heat values. We have already determined that there is a great deal of water loss from the subterranean environment, and the loss of coolant always creates additional heat. An automobile's cooling system is designed to dissipate the amount of heat the engine creates. If you begin to run low on coolant, like your car's radiator, eventually, no matter how cold it is outside, you will overheat.

Cooling and Venting

Water's unique properties have allowed man to use it in both heating and cooling in a wide variety of other ways. Power plants use it to cool their generators, buildings use it to transfer heat, and a welder may use it to cool his iron. Water is used in many ways in our everyday life for more than just drinking, and it's because of this unique ability to cool, heat, and transfer energy efficiently.

As the heat builds below during the summer, it naturally rises toward the surface. As it comes near the surface, it will come in contact with water. The heat will transfer and the water will move it rapidly toward the surface.

Early homes like mine were equipped with gravity feed boilers. This means that the water is heated in the basement and the heat flows up naturally through one set of pipes and into the radiators. As it flows through the radiators, the heat is released and the water is cooled. The cold water is heavier than the hot water, so as the hot water moves up one set of pipes, the cool water moves back to the boiler through a return set of pipes, and circulation naturally occurs. We use water because of its efficiency in transferring heat, and the planet uses this too.

Temperature samples of various deep-rooted vegetation has revealed that the temperature of the leaves exposed to the sun for extended periods of time reflect the same temperature readings found around its base in the shade within one or two degrees. In the forest during the summer, the leaves, trunk, and ground surrounding the trees are the coolest temperature readings of all temperature samples. When the leaf is severed, the temperature of the leaf begins to rise indicating that circulation is taking place, cooling the leaf.

All vegetation is equipped with pores that are called the stomata. These pores open and close in response to many conditions such as the season, the amount of sunlight, temperatures, water levels, humidity, wind, and other factors. During the day, they allow photosynthesis to occur, and at night, they open in a process called transpiration. This is a period of time when science indicates that this process allows for the tree or vegetation to cool down from the daytime's sun. With water flowing through the tree, remaining the same temperature or even cooler than its surroundings, it is clear that the cooling is for something other than the tree.

Each plant is like an individual vane in a large radiator that takes this heat from several feet beneath the surface and allows it to be released at night through water vapor. Just as heat flows up in my gravity feed boiler, so does the heat from the bedrock and subterranean soil transfer its heat through the water where it will rapidly rise to the surface. The surface is warmer than this water, so it remains boxed in below the surface until nightfall when it can be released, and the warmest water will be at the highest level and will be exhausted first.

The amount of release will be directly dependent upon many factors such as the size and type of vegetation, season, temperature, water levels, dormant heat, and relative humidity. The more moisture available within the biome along with higher temperatures will promote more release of water vapor into the atmosphere during the evenings.

This process is a bioengineered marvel. During the summer months, the vegetation alters its stomata for the process of photosynthesis, minimizing the amount water vapor being released into the atmosphere during the day. At night, the pores of the vegetation open and release the heat through water vapor. When fall comes, the leaves turn color and fall off, shutting down this process and sealing the heat in under the thatch in preparation for winter. As winter comes, the energy is slowly released as long as the surface temperatures remain less than the subterranean level. When the season changes to spring, the trees will bud and begin the process over again. If a cold snap occurs for any reason, as we have seen here in Minnesota in the past, the buds and newly formed leaves fall off until the weather becomes warm enough. They will then bud again allowing for venting through transpiration.

Areas inland where the low temperatures decline to 50 degrees at night or less on an overall average, we find the vegetation losing their leaves. Southern locations lose their leaves later and bloom sooner than areas in the north, and as the temperatures rise in the spring, venting begins. A maple tree in northern Minnesota will lose its leaves earlier and bud later than a maple tree in Missouri. Most of South Dakota changes to autumn after most of Iowa, Illinois, and Missouri, and these are all further south, indicating a temperature variable and light levels are both factors. In 2011 and 2012, we had a very warm spring in Minnesota, and even during this warmth, the trees still continued to remain dormant, indicating that the type of vegetation, amount of sunlight, and temperature all have a direct effect upon the vegetation's seasonal alterations.

All nature provides ground cover, even in the deserts with sand. When man converts this land for farming or urban development, this cover is lost and is exposed to the solar radiance, resulting in additional heat gain that drives up the temperatures. Natural cover is a vital mechanism that provides a cool and stable sublayer and aids in maintaining moisture, specifically through dry hot spells. Without this, increased evaporation will rise season after season and the potential for droughts rise year by year.

Both surface and subterranean water aids in cooling the planet by moving much of this heat being boxed in below out of the soil through underground water sources and surface water. Inland lakes offer vast cooling that allows a continual release of built-up heat from below. As surface water levels decline and warm, they become less able to release this built-up energy from below. Areas vacant of surface water resources become more reliant on underground water and heat transfer through the surface vegetation. Man's response to impending drought is to put in place additional wells for irrigation, and in doing this, we also accelerate the heating process by decreasing the subterranean water levels.

When converting forest and prairies into agricultural land, we need to also consider the reduction in carbon dioxide sequestration that will cause a rise in overall carbon dioxide levels. Croplands and grazing land capture far less carbon dioxide than the prairies, forests, or jungles. Utilizing wet crops (such as rice) offers far more cooling and is capable of capturing far greater amounts of carbon dioxide than dry farming due to the life sustained within the water and also increases the cooling factor that the water provides. Such farming techniques have provided wide spread population growth over the years in Asia, but it doesn't have the same effects that nature provides. These areas show significant alterations as a result of their land usage over time.

Seasonal Alterations

If the skies were filled with clouds and the solar energy was drastically reduced upon the Earth's surface, the planet would cool. If there were no clouds and the land was exposed to the solar energy every day, we would experience warming. In this very same way, there are impacts that vary from one season to the next depending upon the season and type of conversion.

When the surface is altered, there are multiple effects that can be measured and they offer both heating and cooling. For example, during summer months, the clouds offer great cooling to the planet's surface during the day, but at night retain the warmth. Forests and prairies have similar warming and cooling effects when compared to farmland, depending upon the season and time of day.

Autumn

During the fall, the forests aid in maintaining warmth until all the vegetation has fallen, and even then they are able to sustain warmth from both the ground clutter and organic material on the surface and the vast rooting system underground. The prairies become layered with years of old vegetation and grass clutter covering the surface, offering an insulating effect for the ecosystem that is driven below ground as the seasons change.

In comparison, cleared fields in the fall are subject to the variable temperatures and can drop down colder in the evenings than the forest and even the prairies. This does aid in offsetting some of the heat gain caused by the clearing of the fields, but overall cooling loss doesn't offset the amount of heat gain. The cropland becomes lifeless and is subject to potential deep freeze until the snow cover aids in insulating it. Even when crops are tilled into the soil, they offer little to no resistance to the cold, have no root system intact, and are absent in comparison to the vast amount of life forms found in the forest and prairies underground. Standing crops offering some shade and are found to be the coldest during this season, while the forests are the warmest.

Winter

Air is an excellent insulator, and we use it to insulate our homes in our storm windows and thermopane glass. In the military, the cold weather Mickey Mouse boots we used in the Marine Corps for cold weather training used air pockets in the rubber to keep our feet warm, even without socks or a liner. Air is used in nature and insulates the subsoil from the freezing cold temperatures above ground to protect life and is found in the thatch. As the snow falls, there are small pockets of air that are naturally trapped by the thatch between the soil surface and the snow. This aids in maintaining warmth and protecting life below while maintaining a fixed level of energy transfer out of the earth. Regardless of the temperatures on the surface, the temperatures remain far more stable and near the freezing point.

Above ground the forests and prairies offer little resistance once covered in snow in comparison to cleared land. Although the forests do offer a wind block, any net gain or loss is negligible during this season. The greatest impact on the planet's heating and cooling is the length of time and the amount of area affected during this season. When this season's length and area declines and is unable to exhaust the summer's heat enough, a gradual increase in warming occurs over the seasons. This creates a carryover effect of heat the following year.

Cropland has virtually no cover and this creates a greater level of energy flow out of the Earth than natural environments. Once covered with snow, the level of energy transfer will become only slightly greater than the forests and prairies.

In the cities and urban developments, we find many homes and businesses that cover the land, and they are all heated. Heated plots of land will seal that portion of land from releasing energy. This will allow the heat to continually rise beneath and force it to vent out the sides of the foundation. Along with the heat from activities (such as vehicles and furnaces), we experience what is known as the urban heat island effect. Minneapolis and the surrounding areas are usually several degrees warmer than the outlying rural areas.

Land cleared for travel such as roads and parking lots during this season remove the snow's insulating effect. Areas where the snow is removed are often subject to deep freeze and is observed in the frequency of frozen water lines.

Spring

The frost in the ground is frozen water and dirt that reaches up to six feet in depth throughout central Minnesota, even deeper as you travel further north. When the snow leaves and the temperatures rise to levels that exceed 32 degrees, the surface begins to thaw. The core continues to drive energy up toward the surface and begins to drive the frost upwards.

In a short period of time, the surface and subterranean heat meet and their temperatures equalize. The axial tilt, atmosphere, and core heat work together to drive the frost out of the ground. If the atmosphere were to do this on its own, it would take several months before it could thaw out regions six feet below the surface. It is the energy from the core that accelerates this process and allows for an earlier spring than the atmosphere and axial tilt can provide on its own.

Spring is the season where we find a vast and widespread alteration that has changed surface weather patterns throughout man's history, often resulting in massive migrations. Saturated with water from the melting snow, nature maintains a cool, moist, and stable environment just above the Earth's surface. This aids in minimizing water runoff as the forests and prairies begin to bloom.

As the forests and prairies blossom and are filled with life, vast amounts of carbon dioxide is sequestered during this part of the cycle and replenishes the air with fresh oxygen. The leaves in the forests and the old grass in the prairies from the previous year are scattered upon the ground and act as an insulator that is easily measured.

The crop fields in contrast are lifeless and barren with no sequestering until well into the growing season, depending upon your location, when the crops become large enough to allow some measurable shade and sequestering. The fields are even absent of ants due to the heat, lack of vegetation, and years of insecticides have aided in this effort. As soon as the snow is gone and the fields are exposed, they begin to rise rapidly in temperature, specifically on sunny days. This turns into dry barren land that only increases the heat and intensity.

Grazing land replaced a vast amount of the prairies and offers very little sequestering in comparison. The soils become compacted and prevent proper infiltration into the sublayers and increase runoff. The prairies vary, depending upon the location, with roots that extended down to 15 feet. Nearly 2/3 of their biomass existed beneath the soil, and this prevented additional runoff becoming saturated during this time of the year. Grazing land is short grass with shallow roots that is trampled and compressed in comparison and offers virtually little to no insulating value.

The forests and prairies take in the melting snows and spring rains becoming saturated and act as a watershed for the planet. This water is valuable, and over the following months, they slowly release this water back into the atmosphere through a process called transpiration. This is vital to areas atmospherically downstream from these locations as it provides them with moisture and continues the water cycle until the water reaches the Atlantic.

The snow and early spring rains can create a sealing effect on the surface soil preventing proper filtration of water into the soil. The farmers till their fields early to allow for the infiltration of water and to aid in preventing this from having an impact on the soil. After planting, large storms and rain can affect the soil with what is known as impact splash, recreating this sealing effect upon the surface crust again early in the growing season.

There is a measurable deviation of ground temperatures that peaks during this cycle. Just as we would preheat an oven, the Earth experiences a preheating effect during this time. Temperature deviations between different soil samples are at their greatest during this season, indicating that this is a preamble for the summer's heat, intensifying this heat that results in a warming effect on the surface. This solar radiance transfers into thermal radiation resulting in a rise in measurable surface temperatures. This heat then increases the evaporation of ground water sources gradually over the years, and after a couple of decades of change becomes known as climate change and global warming.

Summer

As the spring turns into summer, precipitation levels inland begin to decline as the Rocky Mountains complete their spring thaw. The heat continues to rise and the land begins to rely heavily upon ground water sources. The forest canopies and prairie cover remain in full bloom acting as an insulator, specifically on sunny days, maintaining a cooler environment than man-made alterations.

Crop coverage over the land begins to shade enough soil by late-July to begin blocking enough solar energy from the ground to have any effect. By this time in the season, it's too late as the daytime hours begin to decline, having a greater impact than the crop's shade. Then the cycle repeats itself.

In the fall, the croplands' exposure to the elements results in lower surface temperatures. This aids in offsetting some of the increased heat from the spring and summer, but is unable to, over time, compensate for the amount of heat retained, resulting in a methodical rise in surface temperatures.

Atmosphere, Carbon Dioxide, and the Ozone

The atmosphere offers multiple layers of protection from solar influences blocking and protecting the surface. Any resistance to this energy will result in a rise in heat over time; it is one of the thermodynamics' principal laws. When solar energy reaches our planet, it first makes contact with the magnetic field and here we find the first layer of heating in our upper atmosphere. Responsible for blocking harmful incoming radiation, this is also the layer protecting our planet from debris and where we see shooting stars as they burn up entering our atmosphere. Science indicates that although there is shifting in our magnetic poles, its strength and ability to protect the planet has remained unchanged.

Any solar radiation making it through reaches the next layer, the ozone layer. We have all heard about the ozone layer as it was reported to have been depleting during the 1960s and 1970s with our use of fluorocarbons. As we rise in elevation, water's boiling point lowers until the freezing point and boiling point match. I had spent a number of years looking at this characteristic of water, knowing that there was some purpose behind it. One day, I decided to look closely into this and find the matching point and see what existed. I found that this is no coincidence and is a vital characteristic in water that allows for the ozone layer to exist. At a specific pressure, the boiling point and freezing point of water matches, and then it reverses. This means the boiling point is at a lower level than the freezing point, and that below this pressure, which lowers even more as we increase in altitude, water can only exist in the form of vapor and can no longer crystallize or liquefy. When water reaches 0.77 mmHg, which is lower than normal pressure, this occurs. On the average, this is reached at an altitude of 72,980 feet, or 13.8 miles (22.25 km) above the planet, and this is where science indicates the ozone layer begins. This unique characteristic of water allows the ozone to stabilize above this altitude. Ozone, or O3, is a chemical throughout our atmosphere, but above this elevation and at these pressures is far more stable and abundant. The ozone blocks much of our harmful incoming solar radiation, and aids in keeping oxygen and other gases within the lower portion of our atmosphere. Because it offers resistance, it will cause a rise in heat as well, and this is registered above this elevation as the temperatures rise. Water in the form of vapor exists in this area of our atmosphere left over from the creation of our planet and is replenished from time to time through volcanic discharge and from incoming debris from space. Even space shuttle launches contribute to this layer.

The carbon dioxide levels raising and lowering are a part of the planet's cycles. Carbon dioxide has minimal effects during high solar activity, but research indicates it has an effect during the cooling cycles of our planet. Carbon dioxide levels naturally rise during solar maximums by increasing fires and volcanic emissions and decline during periods of cooling, indicating that carbon dioxide affects the climate greater during a lull in solar activity. After these events, they aid in expanding the carbon dioxide in the atmosphere and the impact is felt through periods of cooling.

Eruptions are proportionate with fewer large eruptions than smaller ones. Since 1900, there have been sixty-four

VEI-4 eruptions, nine VEI-5 eruptions, and only three VEI-6 eruptions around the world. Volcanic eruption data indicates that these large eruptions increase during periods of time when heating occurs. So the planet naturally increases carbon dioxide levels after a solar maximum by not only increasing large scale eruptions, but would also naturally increase all activity proportionally around the world.

Around the center of the earth, the atmosphere is swollen due to the heat being applied and remains stable in this region. We know carbon dioxide is a warming agent, but research indicates that its effects are mild in comparison to the impacts caused by land alterations and declining water tables. This would support the theory that carbon dioxide levels would also maintain more warmth during the evening hours and through the winter when the solar radiance declines. Carbon dioxide aids in maintaining this layer in a swollen, heated state along with many other elements including water and methane. None of these heat this layer, but they maintain it in a fine balance and resist outgoing heat, slowing its ascent.

After the 2001 solar maximum peaked, the Earth experienced massive disturbances with events like hurricane Katrina and Rita during the solar minimum years after the solar peak. Swelling of this heat by only four degrees, two degrees north and south will alter and move climates by nearly 140 miles. This means that typical Des Moines weather will now be moved to Minneapolis.

Having a better understanding of the heating and cooling process of our planet aids us in understanding how we have created many of these problems we are experiencing today. Using this understanding, I will share with the reader an activity mankind is engaged in that makes the carbon dioxide argument seem insignificant in comparison. They both do have one common element—fossil fuels. Using Pascal's theory, we were able to bring about some very strong evidences to support a correlation existing between oil production and earthquakes, but there are other far more invasive techniques available today that are occurring right beneath you.

Volumetric Heating/Oil Shale Extraction

In December of 2007, I received a job and dispatched my men to a remote location just outside of Rifle, Colorado. I was informed that this was going to be a backpacking lodge deep in the mountains, so the overnights would be supplied by the business on location and they also offered food on site. When the men got into town, they were informed that they were going to need chains to make it up the mountain pass. The area would lead them off the main road, through a creek, and then up a one-way mountain pass supporting two-way traffic. They were so remote that when they tried to find a radio station, there was nothing available.

During orientation, they found out that this wasn't a backpacking lodge, but an oil petroleum facility. They were instructed to stay on site and any time they were to leave the mountain, they were to be escorted. The company was afraid of protesters finding out about the facility and potentially blocking the pass, so they wanted minimal traffic up and down the mountain. My men called me that evening asking me what I had gotten them into and we got a chuckle out of this, and they did reassure me that they were allowed to leave. Ray told me, "There really isn't anywhere to go anyways." They decided to just stay on site and complete the job as quickly as possible. While working with the other crews on site, they indicated that they had been involved in the construction of many of these sites over the previous year. This would make 2006 the starting date of these facilities.

President Bush had opened up public land for oil exploration, and this was one of the many sites. In order to avoid protesters, they remained as secretive as they could. This new facility would be used to induce radio frequencies, the same technology used in our microwave ovens only on a lower frequency, and heat the shale deep underground and extract the gas, oil, and critical fluids. This process is called volumetric heating. This should not be confused with *fracking*. There are other technologies and experiments being done to process oil shale in order to prevent open mining and be able to maintain an area's ecosystem, but volumetric heating is a new process in massive use today. All underground heating will comply with known physics.

The overall average temperature increase is one degree per 70 feet in depth when we are not near the edges of the tectonic plates. Along the edges of the tectonic plates, there is a great deal of energy exerted against one another, resulting in friction that causes the temperature gradients to be greater near the plate boundaries. Because of this heat and pressure, there are also pockets of oil located along the plate boundaries. Organic material becomes trapped within the layers caused by the subduction of the ocean plates moving under the continental plates. As a result of this heat and pressure, organic material is naturally converted to oil.

The primary oil shale in the Green River area is located at 2,700 feet below the surface. This comes to a rise of 38.5 degrees above the normal surface temperatures, or less than 100 degrees. In preliminary experiments conducted in 2003, they found that the slower they heat the shale, the higher quality of product was attained. Depending upon their heating method, this could take several days when heated on high, months on medium, and up to a few years of heating on low to produce the highest quality. This was tested by applying a three-degree-centigrade rise per hour, per month, and per year increments. The end result over time is a heat level of 325 degrees to 350 degree centigrade (617–662 degrees Fahrenheit) for extraction. A higher heat level is necessary, 400-degree centigrade or more (above

752 degrees Fahrenheit) if extraction is desired within a few months.[3]

The crews installing this site had been installing them throughout the previous year in areas of Colorado, Utah, and Wyoming. This heat is more than 500 degrees above the normal temperature, and because the core of the Earth generates its own heat that comes up through the surface, it is only a matter of time before this heat permeates to the surface. As this heat continues, it will act as a thermal shutoff for the land beneath this area, boxing in the heat below and driving both temperatures and pressure up below the area being heated. The expected results of such widespread heating would be a rise in surface temperatures proportional to the heat applied as well as the energy dissipated into the surrounding environment. Physics tells us that *all* the energy will come to the surface in time and be felt greatest at the source. What this means is if this process ended today, the heat would continue to permeate for years to come until all the energy that was applied fully dissipates.

Additionally, these temperatures would naturally reduce ground water sources at all levels including subterranean cooling. Altering subterranean cooling has a ripple effect downstream by carrying this energy away from the heat source to downstream locations through underground water. The water basins will determine the directional flow of this heat. Oil shale in Alberta will travel north to the

Arctic and up through the Mackenzie River and thaw the permafrost from the bottom up.

Utah, Colorado, and Wyoming sits on the continental divide and this heat will travel south and both east and west. Because of the location being on the easterly edge of the Rockies, the effects will be felt greatest in the prairies of Colorado, Kansas, Oklahoma, Texas, and New Mexico first and travel westerly, but at a slower pace due to the vast mountain range affecting Utah, Arizona, Nevada, and California.

Although the planet's heating and the impacts from the sun have a far greater effect than this process being induced by man, the sun is not capable of heating this layer, or the surface, to these temperatures. The *only* natural process known that can alter the temperature gradients, and to these depths, is internal volcanic activity. Additionally, this heat is continually building throughout the year and is not allowed any time to release the energy it naturally achieves through the tilting axis. Not very far north from this location, we find Yellowstone. This process is not limited to North America and would also alter other regions in the world using similar technologies, heating the planet to extract elements.

Based upon thermodynamic laws and principles used to determine the boxing in effect of heat, this process only continues to reaffirm this effect upon the Earth. Based upon this law, as long as heat is injected and maintained, all the layers beneath it will continue to rise and build until it is able to breach the heated portion and to the surface. Increasing heat will cause expansion and pressure to build, along with the forces of the weight of the Earth above, the gas and oil can then be extracted.

Although mankind takes great pride in this current technology, there are known physics at work that haven't been fully considered. Previous oil shale technology once required massive surface alterations and moving massive amounts of Earth for processing. This current technology leaves the current ecosystem intact and the processing is done in place. Our ability to accomplish this has created a very impressive local oil market with a great deal of wealth and greed, but consequences are inevitable.

The process was tested in 2003 with the results, indicating that best results occurred after heating was applied for three years. This process began in 2006, when three years are added, we come to 2009 and the North Dakota oil rush. Coincidence? It was not an accident, and this heat that is being injected is also the basic principle the planet uses as a thermal shutoff valve trapping the heat rising up from the core. The heated shale slowly liquefies, expands, and then increases in pressure and becomes oil and gas. Heat and temperatures require time as the variable and after three years of building, it is ready for extraction.

The Pacific Ocean brings its moisture into the Rocky Mountains throughout the winter months and the cold air compresses the atmosphere and forces it to build on the western edge of the Rockies. In the spring, there is a slow methodical thaw every year that creates what is known as our North American Monsoon season. This is a normal yearly process that brings life-giving rains easterly into the prairies and forests.

Ground water evaporation is inevitable with temperatures exceeding 616 degrees. This will cause a rapid thaw in the spring, resulting in early spot flooding across the plains, and this is what we have experienced the past several years. When the land is unfrozen, it rapidly becomes arid and dry due to the quick thaw and latent heat below. Any water that comes in from the Pacific will now be used up by the natural resources. As a result, the air becomes vacant of water vapor causing late summer droughts that extends across the prairies. Our droughts, as confirmed previously, are not a result of a lack of water during the course of a year, but how the precipitation is being distributed throughout the year.

Today, we are beginning to see alterations in our normal weather patterns as a result of this technology, an expected side effect of heating. This accounts for Minnesota's unusually late drought conditions that have occurred in 2011–2012 seasons. One began on August 17, 2011, and the next year on August 16, 2012. Here is a graph of recent precipitation levels:

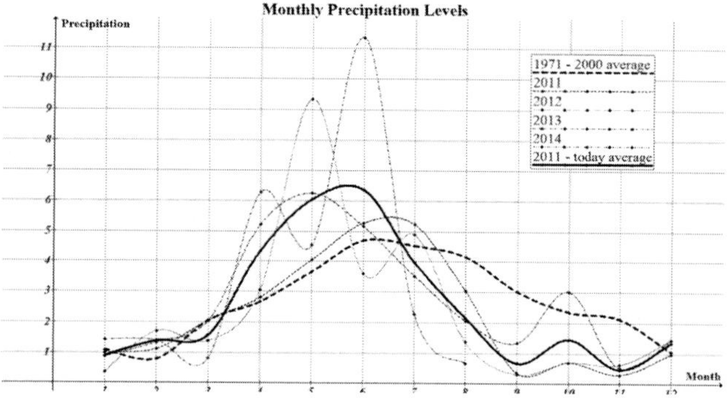

In this graph, we can clearly see that there are serious alterations in precipitation levels that have developed. Early spring rains and lack of sunlight has hampered early planting while late summer drought conditions have stressed many areas across the Midwest.

Heating means expansion along with a rise in pressure and temperature over time. The potential for earthquakes and volcanic activity rises proportionally and increases with time. Additional side effects are rapidly declining water levels and alterations in the typical seasonal rains. With all of these, there are ripple effects that come such as wildfires, intensified storms, and declining crop yields.

Other documented areas in the world utilizing this technology are China, Estonia, and Brazil with resources available around the world for this form of extraction.

When we go after any elements in the ground and use heat to extract them, there are known physics at work with expected results. As long as there is a layer that remains even one degree higher than the layer beneath it, the heat will continue to build exponentially until it breaches the applied heat. If this heat is not allowed to exhaust as the planet does every year, there will be serious ramifications.

The Earth uses 36,000 feet of rock and topsoil to insulate this level of heat, so nothing man can do can alter this effect as long as they remain turned on. If turned off, it will take several years for the heat to exasperate from these affected areas. During this process, earthquakes would naturally rise due to displacement.

This process that is being used to continue to drive our economy on gas and oil is having a far greater impact on increasing temperatures at a far greater rate than carbon dioxide levels could ever warm our planet and is one of the greatest threats today, specifically to people who reside on the North American continent. It would be ideal to move away from fossil fuels for both of these reasons, but this is what is driving the economy today, so there are many industries and nations that will present great opposition to such a decision. This specific process of extracting oil and gas is having a direct effect upon our heat gradients, causing a rapid rise in surface temperatures that will result in a great catastrophe in time. Once again, time is of the essence.

Electron Flow and Magnetic Poles

In researching heat and resistance, one of the layers I researched was the magnetic field. This field, being generated by the differentiation at the poles, reaches deep out into space at the midlatitudes acting as an invisible force field for our planet.

I began to think of the electricity in my home. When we use anything electrical, we need to hook up two wires to complete the circuit, for example, jump-starting our car. All electrical cords have at least two wires, but one is connected to the ground back in the panel. All of our AC power comes in on one wire per phase, and then has a return path into the ground, indicating current flow and conduction of the Earth attracting electron particles. I have been battling Earth grounds since I began as a service technician in the mid-1980s. Ground has different potentials that can vary from a building to another source such as another building, light pole, or a gate. When this occurs, especially with cameras, you will get feedback and with CCTV cameras, this is often seen as a hum bar across the screen. A typical voltage variance seen ranges from 0.1 to 1 volt AC differential. This can be easily resolved by installing a ground fault isolation transformer to physically separate the grounds.

This indicates that the ground itself absorbs or attracts electron flow and is found deep in the depths of the magma beneath us. The centrifugal force of the planet's rotation would exert a great amount of energy outwards at the midlatitudes in comparison to the poles. So it is in the midlatitude that we find this energy field at its peak. This magma, driven by the same forces moving the tides on the surface, is also moving this magma. Around the midlatitudes, the heat rises substantially as we move toward the equator. This is due to the forces being applied from its rotation and amplified by the gravitational pull placed upon it in conjunction with the amount of sunlight exposed to the Earth. This causes massive heating at ground level causing rapid evaporation, storms, and lightning. As this lightning stimulates current flow in the magma beneath the surface this creates an electric field around the Earth. This electronic field around an iron and nickel core creates and generates our magnetic field.

> Ours is, without hyperbole, a dynamic planet. The flow of liquid iron in Earth's core creates electric currents, which in turn create the magnetic field.[4]

A magnetic field has specific physical properties associated with it. In electricity, this is called the right hand rule theory that we use in electron flow. If you know the electron flow, then using your right hand, you place your fingers in that direction. In this case, the force is west to east so your palm would be on California while your fingers reach the East Coast. Your thumb then indicates the North Pole, and this is where it is and complies with electromagnet principles.

As the Earth heats, storm activity increases, causing the water to be drawn up, cooled, and then return back to Earth. This offers shade, cooling, and life-sustaining rains to promote growth. The storms increase lightning increasing electron flow, resulting in an increased strength of the magnetic field. Increasing the magnetic field's strength also increases its resistance to incoming solar radiation, suppressing more of the solar energy. So we can see that as the heat rises, the field strengthens and as it cools, it declines. This creates a bioengineered, electromagnetic, protective field that automatically adjusts according to the planet's heat variables.

As it cools, there are fewer storms and more sun, which creates more radiance upon the planet's surface. This creates a declining electrical field resulting in a decline in the Earth's magnetic field that creates a warming effect. The harmony that exists is far greater than I had ever thought. These actions and how they work are similar to a bioengineered form of a machine with checks and balances beyond my imagination.

Soon I was asked about science and the specific discoveries that oppose this where they indicate pole reversals in rocks and minerals. Here, we need to look toward Antarctica to understand plate movements. Antarctica was once near the equator according to science, and I also concur with these findings. This means the entire continent at one time had specific embedded pole signatures, but today at least half of the continent would show a reversal from its original location. Continents move, twist, roll, rise, and sink based upon the many forces at work, and water continually moves and covers up many of the Earth's hidden secrets. Our planet is very dynamic and the plates are constantly moving with multiple forces at work.

Tectonic Plate Movement

The tectonic plates have multiple forces moving them and are very dynamic. The centrifugal and gravitational forces are the primary contributors by maintaining the seas and magma in constant movement. Following the laws of expansion and contraction, every year as the axis shifts, the continents expand and then contract. The deep oceans, regardless of their location, maintain a semi-constant temperature below 40 degrees and are not directly affected by this shifting heat. During this process, the oceans and plates will interact with one another causing shifting and movement. Like a snail, the plates expand and contract every year with the gravitational forces, ocean tides, and neighboring plates, creating a force that attempts to slowly move this land.

The oceans offer great influence on the plates, applying force from west to east across the surface. The force of this power can be seen as the Pacific blew out the land bridge that once connected South America to Antarctica and left much of its remnants in the Atlantic.

Just a very short distance beneath the surface of our planet, there is a very dynamic planet at work. The magma we see coming up from volcanic activity demonstrates the amount of energy and heat that lay beneath the surface. As this magma reaches the surface and the pressures decline, the boiling point also declines creating a liquefied state as it comes to the surface. Under the pressures of the Earth, it will act like other elements and the boiling point rises.

The heat from the core in a spinning sphere will be propelled outwards and will be the greatest at the midlatitudes. This heat, like the oceans, will move this magma and is picked up as deep level earthquakes beneath the surface. This magma from the northern and southern hemisphere will come together in the midlatitudes. The heat from the core will be exerted out and will create a form of an El Niño and La Niña events beneath the surface as we see on the surface. This heat generates movement and energy from the center of the Earth outwards, applying force upon the plates from the equator toward the poles. When cooled contraction takes place, the energy will draw the landmasses from the poles toward the center of the planet.

Based upon the concept of a megacontinent at one time existing near the equator, such a landmass so close to the equator would have been split apart by volcanic eruptions and earthquakes based upon the planet's current heat budget. The planet, through millions of years, has adjusted the plates based upon these principles of heating and cooling to obtain an optimum balance. Just as the southern hemisphere slowly moved Antarctica to the bottom of the world and covered it with ice, there was a balance found. The principles of expansion and contraction explain the movements of our tectonic plates across the surface of our planet.

When researching earthquakes above 7.0 since 1900, there was also a very solid pattern around the world. Earthquake depths were researched, and it was found that there is an average of eight to ten earthquakes of this magnitude since 1900 and various depths until 34.5 to 35 km in depth is reached. At this depth and in this thin layer, we see 169 earthquakes around the world. In comparison, there is only eight occurring in the range of 35.1 to 36 km, eight from 33 to 34 km. Universally, around the world, there is a defined layer that exists at this depth where the plates are in constant movement with one another. I decided to further my research into the effects of heat on the plates.

Earthquakes and Heat

Whenever heat rises, there is more pressure and energy resulting in an expected increase in earthquake activity. Earthquakes will proportionately increase in their magnitudes over time, meaning that for every ten level-3 earthquakes experienced, a level-4 will occur, and for every ten level-4 earthquakes generated, a level-5 can be expected.

Because of the massive amount of earthquakes experienced around the world, I decided to look at earthquakes at a level of 7.0 and larger and limiting the depth to 35 km since 1900. This is the barrier where the heat rises high enough to melt granite at sea level. This is the lower layer of the crust that would be affected the greatest by the expansion and contraction of the earth's plates. Beneath this area, which has not been researched yet, should be a more constant and even heat flow.

This research data continued to demonstrate the heat gradients around the world and how the plates act. Heat creates pressure and expansion that will result in greater levels of earthquake activity. In a perfect sphere and according to our orbit, there should be more earthquakes in the midlatitude of our planet due to the centrifugal force of the rotation. Additionally, the southern region should indicate a greater number of earthquakes due to the planet's orbit, bringing our planet closer to the sun during their summer. The other reason earthquakes will be greater around the midlatitude is due to the amount of area, due to the larger diameter in the middle of a sphere compared to the other regions. Although an influence, the following figures indicate that heat gradients and their alterations on the exposed plates have a greater influence than the amount of area heated.

Breaking the planet down, there are 180 degrees of latitude in our world. Separating these into 18 increments,

10 degrees per increment, large-scale earthquakes above 7.0 were then plotted since 1900.

From the equator to the South Pole, the numbers are as follows: 296, 221, 168, 55, 26, 28, 19, 0, and 0. As we travel toward the South Pole, the number of earthquakes gradually declines other than the 28 being greater than the 26.

In the Northern Hemisphere, from the equator to the North Pole, the numbers are as follows: 162, 167, 125, 190, 170, 118, 13, 2, and 0.

Between the thirtieth and fortieth parallel north, there are 190 earthquakes, and in this region, we find massive land distribution with more earthquakes than any other in the north. Compared to the southern region, there are only fifty-five. This is followed by the next abundant landmass between the fortieth and fiftieth parallel with 170. In the southern region, we find that there are only twenty-six. There are more earthquakes along this latitude than even the area from the equator just north to the tenth parallel, and there is significantly more land available too. Between these latitudes in the north, we find the main continental plates of Asia, Europe, and North America. In comparison, the southern regions have massive bodies of water.

This clearly indicates that heat distribution across the plates exposed above the water have the greatest influence on earthquake activity.

Understanding these forces, earthquakes were reviewed once again to see if any pattern could now be ascertained. This is also the area of research that I had to discontinue in order to write this book, so I will continue to research this area after publication. Although not complete, there are many findings that are substantial that only continue to support the planet's operation. The heating and cooling, causing both expansion and contraction of the plates, are examined by seasons according to their locations. During periods of transitions between hot and cold, expansion and contraction, the surface demonstrated heightened activity in this layer. Heat causes pressure and stress, and earthquakes are a form of stress cracks, the same we find in machines subject to extensive heat.

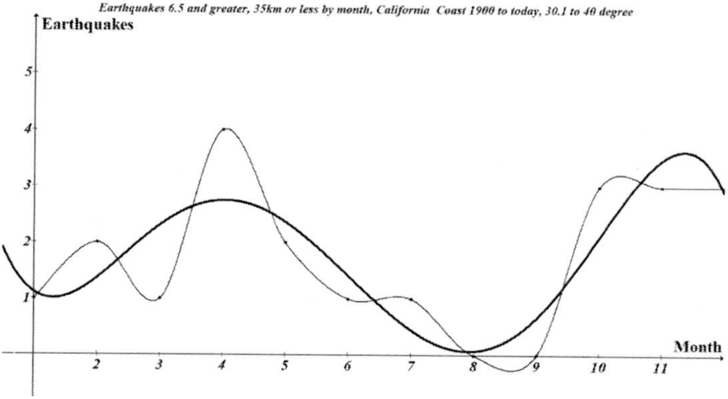

In this graph, we can see that the West Coast of the US has never experienced an earthquake of 6.5 or greater in the months of August and September over the past 114 years. This is not a coincidence and such historical activity occurs around the world with different variables depending upon their location. In California, they actually have experienced a period of 86 days, July 21 through October 15, with no earthquakes at 6.5 or greater since 1900.

Around the world, this pattern exists and as we travel further away from the equator where the temperature gradients alter greatly, the more evident these patterns appear. These spikes are six months apart, and in California, this is in April and October. As we travel further from the equator, they alter slightly into May and November due to the axial shift.

As the earth heats in the spring, the laws of expansion begins. As the landmass expands, it applies pressure with neighboring plates, and this pressure seals off volcanic venting and allows the pressure to build. The plates are heated and fully expanded in late July and maintain expanded through September. In October, the planet has shifted far enough to begin cooling and contraction of the plates begins. As they contract, they also open volcanic vents and fissures that promote eruptions. These eruptions also generate earthquakes throughout this cycle until heating and expansion begins again in the spring. Volcanic activity explains the increased activity during the winter months in comparison to the summer months.

As the plates expand and shut down the volcanic fissures along the tectonic plates, there will be limits to the pressure it will be able to maintain. We mimic this operation in the automobile's cooling system. The automobile has a radiator cap that maintains pressure when heated with a relief valve as a safety device, preventing the cooling system from over pressurizing, and so does the planet. The pressure will build and begin to ooze out, just as we are seeing above ground in Hawaii and is creating new islands off the coast of Japan. Deepwater volcanic activity will naturally increase during times of heating, and this will take place completely out of sight. The Mid-Atlantic Ridge and areas along the Pacific's Ring of Fire will naturally rise, and there is nothing that can raise the sea levels faster than the process of water displacement by increased lava flow. This expulsion of lava will not only displace the water but also cause the water temperatures to rise, altering the jet streams and precipitation levels.

Antarctica has been cooling while the North Pole has been declining with the exception of the peninsula that extends out toward the South American continent. There has been drastic ice loss in this area, and it is also an extension of the Mid-Atlantic Ridge in the southern Atlantic Ocean. With the Arctic nearly landlocked and its major boundary of water is with the Atlantic Ocean, there is clear indication that the Atlantic is warming at a faster pace than the rest of the world's oceans.

If the plates do not have significant time to cool and remain expanded through the winter months, the danger rises substantially of a very large and catastrophic eruption. Along with this pressure and heat would be an expected rise in the amount of earthquakes.

I am looking forward to furthering my research in this area and should have this complete by publication. As usual, this will then move me into another direction of research.

EARTHQUAKES, LAYERS, FRICTION, AND KINETIC ENERGY

Driven to continue with my research I was able to complete this section before print. As a last minute addition to this book, I am now able to share this information.

Earthquakes above 7.0 were examined since 1900 at various depths using the USGS website.* There have been many modifications and technical advancements over the years allowing for greater coverage and accuracy. Since 1900, the world has experienced a total of 1,367 recorded 7.0 earthquakes and greater. From the surface to 35 kilometers in depth (21.9 miles), there have been 990 earthquakes, or more than 72%. The continental plates are in constant movement shifting and altering, then generating earthquakes at various levels as they impact each other, and this is where the most damage occurs, near the surface.

---

* http://earthquake.usgs.gov/earthquakes/search/

In this graph earthquakes were plotted according to the following. From ground level to 5 km were plotted at 5, from 5.1 to 10 they were plotted at 10 and so on.

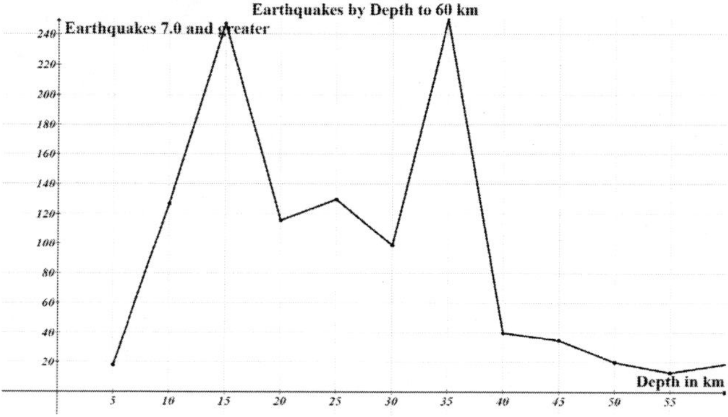

When researching these earthquakes in 5-kilometer increments there is a clear division located between 10.1 and 15 kilometers deep and another 30.1 to 35 kilometers beneath the surface. At 15 kilometers we see the impacts between the surface plates as a result of seasonal expansion and contraction along with tidal movements and accounts for 247 earthquakes, or 18% of all earthquakes.

What lies beneath this layer is truly unknown. The deepest we have drilled to date is 12,262 meters in the Kola Peninsula in Russia. Throughout drilling, scientists were surprised at their findings including plankton microfossils at a depth of 6.7 kilometers.* The Kola Peninsula is located in the far north along with Sweden, Norway and Finland. At this depth, and this far north, the temperature was 356 degrees. We were still unable to pass through the very first transitional layer located at 15,000 meters below the surface as the borehole began to fill in on itself due to the heat and pressure. In comparison, the Mariana Trench is less than 40 degrees and is near the same depth.

From 30.1 to 35 kilometers there have been 251 earthquakes and account for more than 18% of all earthquakes. Within this 10-kilometer region, 30.1 to 35 and 10.1 to 15 kilometers, 36% of the world's most destructive earthquakes occur. From 35.1 to 40 kilometers there are only 40, more than 6 times less, and continues to decline as we travel deeper. From 50.1 to 100 kilometers there were 81 earthquakes. These are transitional layers with the 10.1 to 15 kilometer range being the upper level of the oceanic plates coming in contact with the continental land plates. From 15 kilometers to 30 kilometers, where earthquakes decline, indicates a harder layer that would represent the tectonic plates girth. The 30.1 to 35 kilometer active transitional region is the contact points where the lower edge of the tectonic plates come in contact with each other and are driven down by pressure into the more rigid region below.

---

* http://www.damninteresting.com/the-deepest-hole/

Earthquakes were then plotted out around the world and divided by depths in 50-kilometer increments. In a perfect solid sphere the earthquakes should decrease proportionately as we increase in depth due to reduction in surface area of a sphere. If a layer rises in earthquake levels, then this would indicate a transition layer as we found at the 15 and 35-kilometer depth.

From 50 to 450 kilometers earthquakes did slowly decline as we increased in depth. There was one layer between 350 and 400 kilometers that indicated a slight rise but with so few earthquakes a separate graph was ran using earthquakes above 6.5 for closer observation. The following graph reveals that there is another transitional layer that mirrors the crust. One minor transitional layer that exists between 350 to 400 kilometers in depth, (219 to 250 miles) and the other, a major transitional layer, located at a depth of 550 to 600 kilometers (343 to 406 miles). There have been 53 level 7.0 and greater earthquakes from 550 to 650 kilometers deep. In this 100 kilometers there were more earthquakes than the preceding 300 kilometers combined, 48. Just as the atmosphere has layers, there are also layers existing beneath the surface demonstrating how dynamic our planet truly is.

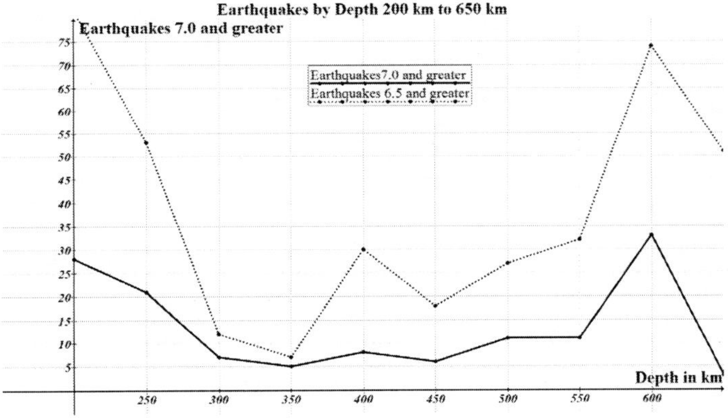

In this graph, the total amount of earthquakes between 350.1 and 400 kilometers in depth is plotted at 400. From 400.1 to 450 is plotted at 450 and so on.

The Earth is sealed and temperatures will rise as we increase in depth due to the increased pressure. Earthquakes indicate movement, and movement means friction that will result in increased heat due to the kinetic energy created. This is the kind of heat that results from rubbing your hands together when they are cold, or starting a fire from rubbing sticks together. Under this pressure, movement will increase heat flow towards the surface.

Below the surface there are points of friction throughout the layers as earthquakes demonstrate, but areas that demonstrate a rise in earthquakes will also have increased movement resulting in a rise in the amount of kinetic energy. At the 35-kilometer and 600-kilometer depth (375 miles) we find transitional layers where friction points provide massive amounts of energy under a great amount of pressure that will slowly rise towards the surface. This supplies heat to the surface, and this is separate from the core.

Between the layers and along the borders of the tectonic plates, heat will be generated by friction resulting in a softening of material along plate boundaries. Created by the massive shifting of weight generated by the tides and the seasonal expansion and contraction of the continental plates, the plates maintain in continual movement from this energy and heat becomes the result.

Science has often made attempts in estimating the temperature of the core, but according to this research the core may be less extreme than originally thought. This evidence clearly indicates that friction between the layers provides a continual supply of regenerating heat separate from the core. These layers are fueled by the friction and pressure from above and they gradually erode the lower contact point of the plates. At the surface we see this movement from this activity in the formation of islands and increased landmass from volcanic discharge.

We have always assumed that all the heat rising to the surface is coming up from the core, but this brings into the equation the amount of energy being generated from friction. Based upon known thermodynamic laws, friction is kinetic energy that produces heat. Although the amount of heat and pressure in these areas are beyond our comprehension, in order to determine the temperature of the core, we would have to reduce this energy from our equation to get an accurate measurement.

Just as the temperature gradients exist in our atmosphere, there are also temperature gradients beneath the surface with different resistant levels. To what degree is unknown, and we may never know due to both the heat and depth. Our perception of our planets core heat mirrors our early perception of our atmosphere, a gradual outflow of heat. As time has gone by, and as we have entered into space, we have come to realize there are temperature gradients above the earth, and they also naturally exist below.

In order to get a better understanding of these layers I decided to try to estimate the amount of activity. From 1900 to today, March 7, 2015 there have been 42,070 days, give or take a day. Using this figure we can determine that there are just over 1 million hours, or roughly 60 million minutes.

Researching earthquakes from the surface to 35 kilometers there have been 990 level-7's, 6,909 level-6's, and 40,673 level-5's. The overall average for earthquakes near the surface at these layers is six to one, for every six level-5 earthquakes, a level-6 occurs.

Using these calculations and multiplying them down to a level-1 earthquake and adding them all together we can determine that there is a level-1 earthquake or larger occurring in this layer every minute on an overall average since 1900. There are eight level-2's and more than one level-3 occurring every hour, and one level-5 occurring every day. This is a great deal of movement occurring under a very extreme high-pressurized environment, and is increasing at a dramatic rate today.

Calculating earthquakes in the deep transitional layer of 550 to 650 kilometers the same overall multiple applied, a six to one ratio. Less prone to earthquakes, and under enormous pressure, there is one every forty-five minutes, and a level-4 occurring nearly every week. The amount of energy needed to register any earthquake more than 343 miles beneath the surface requires an extreme amount of energy.

With the exception of one level-7 earthquake in the Straits of Gibraltar, all of these extreme deep-level earthquakes have occurred in the Pacific Rim either in the Far East, or the Pacific coastal ridge of South America. The Pacific Ocean is the largest ocean and has greater movement from tidal forces than any other ocean, but this alone doesn't account for this differentiation. On the other side of the world, the Atlantic Ocean has a ridge that allows for pressure relief and has never had a deep level earthquake of this magnitude, and the Pacific is vacant of any venting with the exception of volcanic activity. This gives us some indication of the size and magnitude of this ridge extending down to extreme depths in the Atlantic.

From 35.1 to 350 kilometers is what we would expect to see in a more rigid mass slowly dissipating heat with a continuous reduction in earthquakes as we move deeper. This would be a characteristic of a more consistent and semi-rigid layer. Depending upon the temperature gradient and the material present in this layer, this layer could form into a more stable region and would explain this. As pressure increases, temperatures rise and the boiling points increase. If the temperature gradients slowly decreased in comparison to the surface, and the pressure increased high enough to raise the boiling points faster than the layer could melt, then this area could become more rigid than the layer preceding it. As the Kola bore hole demonstrated, we know very little about what lies deep beneath the Earth and is in need of further research.

After 700 kilometers, 438 miles, earthquakes subside. The deepest ever-recorded earthquake was 888 kilometers in Panama, 550 miles beneath the surface. This is only 14% of the Earth and there is still another 3,409 miles to the center of the Earth.

As heat rises due to alterations on the surface, earthquakes rise in activity due to the differentiations in pressure caused from the expansion and contraction of the Continental plates. This creates a feedback mechanism increasing the energy flow deep beneath the surface. As time continues, and both heat and pressure build, the amount of magma flow, eruptions, and earthquakes will naturally rise. As the planet cools, as it did during the 1940's to 1970, the opposite was true and they declined.

Earth's Historical CO2 and Ice Age Events

The following graph[5] is the historical CO2 levels of our planet. Based upon this research, high CO2 levels occur when surface vegetation is decimated. Within this graph, there is a cycle of events very similar to solar cycles. Every 100,000 years, some event is triggered that affects and alters the surface of our planet. Although human activity will produce such levels, the chances of him creating such an event every 100,000 over the past 800,000 years is highly unlikely. Like clockwork, these events occur, and there are three known events that could produce these results.

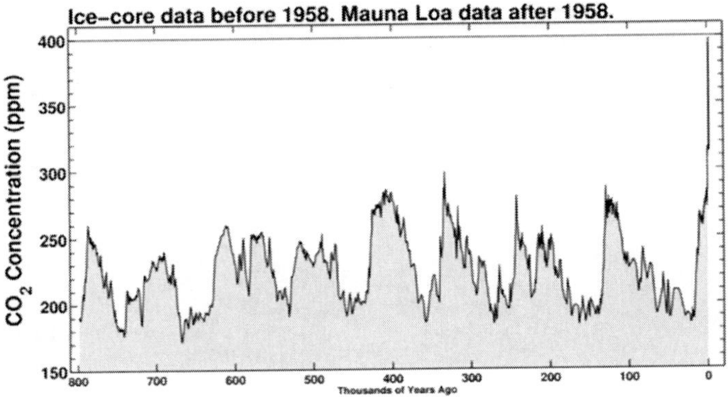

If the sun or the Earth's core were to have a 100,000-year cycle of cooling that in some way regenerated itself, this would explain such an event as seen in this graph.

This research indicates that the planet has the ability to not only protect, but has the ability to repair itself too. It's self-generating and self-correcting. Science indicates that an impact occurred, triggering the last ice age and wiped out nearly everything upon the planet. Based upon the planet's protective and regenerative abilities, it is highly unlikely that one impact would trigger such a long-lasting event.

Based upon the timing of these cycles and scientific data, it appears that our solar system has a high probability of intersecting with another system or debris field somewhere out in space every 100,000 years. Such an event would explain not only the timing in this graph, but also the long-term impacts and devastation. By encountering such a system, our planet could be subject to multiple impacts for a period of time causing long-term cooling and would result in a wide spread extermination event. Thankfully for us and according to this graph, such an event has just passed us.

Weather Modification

Weather modification is the manipulation of the weather through artificial means to produce desired results and has been experimented with since the 1950s. The most common form that we have all heard about is cloud seeding. This process is used in an attempt to bring rains to areas in need of water, or to bring snow into ski slopes for early business. Modifying the weather is not a new process, and over the years, research has expanded further to calm potentially damaging weather such as tornados or hurricanes. It has also been explored and used to aid the military in specific operations or for tactical purposes. One such report available is called "Weather as a Force Multiplier: Owning the Weather in 2025"[6] indicates the military's perspective and current views regarding weather manipulation. What ever the reason or purpose, if the government has the abilities to invoke changes and improve crop production, we would not only want them to be involved, we would expect them to be. Based upon this knowledge and history, it is very clear that our government is actively involved in weather modification and has even admitted to some form of activity. So the question remains, exactly what and how are they accomplishing this?

Using historical data and understanding the cooling effect that air pollution had on our planet between the 1940s and the 1970s, research for an alternative to pollution has been explored as a blocking and/or reflecting agent for the incoming solar radiation. There are many patents available, even huge umbrellas in space have been reviewed as potential cooling devices to combat climate change. In the very beginning of my research, I indicated that implementing such a program would offer a cooling effect upon the surface over time. What man had accidentally done in the past could now be implemented with direction today to counteract warming trends.

My early research indicated that this needed to be a two-fold project of dimming the incoming solar radiation while replanting the earth. Not doing so would mean a continual dimming of solar energy upon the planet, and these processes would need to be a part of our daily lives forever, and this is an unacceptable response for the sake of our children.

Clouds are naturally developed when water vapor rises and attaches itself to a particulate where it then transforms into a liquid state. When enough come together, we begin to see them as clouds. Particulates accumulate in the atmosphere from many forms of natural activity such as dust storms from the Sahara, California wildfires, or the pounding ocean surf. An atmosphere saturated with water vapor but absent of particulates will not develop any cloud formations. Understanding the physics behind this, adding the proper particulates into an atmosphere that is saturated with water will develop clouds.

Cloud cover offers both heating and cooling effects depending upon the season and the time of year. Hypothetically, if the world became fully clouded, we would cool, and if there were no clouds, we would gradually heat. Understanding this supports the theory that increasing clouds will decrease the heat.

In a NASA report in 2003, the following was addressed. After airliners were grounded in September of 2001 attack, many questions began to rise about the influence of contrails on the skies, specifically weather. This report makes many indications that drew me to researching this much closer. The purpose behind the report indicated that, "The results of the analyses should be valuable for improving the parameterization of contrails in climate models."[7] And it was in their conclusion where they confirmed this potential exists when they wrote, "Contrails are a source of anthropogenic cloudiness. Similar in physical properties to natural cirrus, contrails affect the atmospheric radiation budget and may influence climate."

Geo-Engineering

Geo-Engineering is a specific science that focuses their efforts on combating climate change. Atmospheric obscuration is the use of technology and available resources to counteract global warming by blocking and/or reflecting the incoming solar radiance. Governments around the world have remained silent to these practices, and secrecy leads to conspiracy theories. Combating global warming through atmospheric obscuration could lead to potential liabilities, and it's in these liabilities that would create the level of secrecy that is being experienced today among the world leaders. Conspiracy theories rise out of fear of the unknown, and then speculate the worst and this is one area that doesn't seem to have any shortages in conspiracy theories.

Contrails have been a part of our lives since we have all been young. Living in Minnesota, we understand the influence of temperatures on exhaust smoke from our cars as they bellow out smoke in the cold winter air. My neighbor's father worked out at the airport as a mechanic on the jets, and we would frequently lay out in the yard and watch the jets flying overhead while he would identify them. The highflying jets, specifically the B-52s, would often leave long trails behind them that I thought was pollution, but sometimes they weren't there. I remember his father explaining that it was actually water that was freezing and becoming ice from the exhaust. He told us that some water comes out of those jet engines, and they are very hot but the air is very cold and what we see are ice crystals. When we asked why they disappeared, he told us that the sun heated the ice and they quickly evaporate. It was a simple explanation for a young boy and there is more to it than this, but it is also the basics of understanding them. On some days, they seemed to last forever, but when I would turn around to look at them a little while later, they were gone. By watching them, I did understand that sometimes they could persist for quite some time.

The word "Chem-Trail" is often used when referring to aerial obscuration, but there is also bountiful conspiracies attached with this word. I hope to clarify and put many minds at ease by not only explaining these trails, but what they are so we can stop worrying about what they aren't.

More importantly, we can explore the known physics and impacts to our world and our climate.

After first hearing and seeing these trails, the first question I had to ask was, "How long have they been doing this?" Most people don't notice them or watch the skies because contrails are expected, even I hadn't since I was a kid. I then became very concerned about their potential impacts, specifically for the health and well-being of my grandkids. I did some research, talked with many people about them, and began to rethink these trails, after all, these were the types of trails I was recommending when I first started this research, so I decided to begin tracking them.

My coworker and I began by judging the distance of the planes and contrails to see how far we could observe them before they disappeared into the horizon with the naked eye. We had determined that on an average day, we could see them up to 50 miles ahead of us before their size obscured them to the human eye.

To convert a spherical observation as we see outside, you multiply the square root of the radius by pi. (50 × 50) 3.1416 equals 7,854 square miles. This is the amount of sky above the earth that one of these planes can be seen in a given area at any one time from level ground.

The first thing we noticed was the unique and persistent trails behind specific aircraft. Even when there were noticeably long contrails, they would begin producing a permanent trail, and then return to contrails again after traveling across the sky. Skeptics claimed these were caused from wind shear, but one after another at different and same altitudes, different directions, crossing paths, it became clearer that it wasn't anything in the air but on board these planes. Their reasoning just didn't make any sense. Wind shear was very obvious in many of the trails we monitored, making them look as if they would appear, disappear, and then reappear again. These were different; a new and unique trail that resembled the release of a fogging agent like the machines used to fog for mosquitoes, but it never disappeared but expanded over time. I have traveled and followed them over 100 miles before they would become so large that I would lose them into a haze of other masses, creating a vast hazing effect atmospherically downstream.

I downloaded an application for my tablet where I could get live air traffic feeds with a slight delay. We tested the application at the International Airport and found about 70% of the air traffic was responding with this application. Other aircraft in the sky that would not be registered would be private flights such as a senator, the president, or possibly a celebrity. Military flights do not show up on any applications either.

As we continued to observe these flights over time, we became good at finding them, plotting our location, and predicting when and where they would be in the sky. One by one, we saw these specific aircraft with these trails, and not one would show up on the application. We continued throughout the days and weeks, not a single one showed up on the application. My attitude shifted, I wanted to be proven wrong and find one flight that was leaving these specific trails, and I never found one. As long as I had observed these trails up to today, I have never found a jet on the air traffic applications available that were leaving these specific trails.

Some theories indicate that airliners are spraying the skies. My research does not concur with this. I did come across one patent indicating this was researched but was found to have a warming effect and was discontinued. I have not been able to locate this document, but research indicates that these are not commercial flights and are solely unidentified flights.

I then decided to take some rainwater and have it tested shortly after seeing massive amounts of these trails in the sky. The results came back with an unusual spike in magnesium. Magnesium is an element found naturally abundant near the oceans. The further inland one travels, the less magnesium there is. Magnesium is also the same element in my kid's inhalers to aid in their breathing. If this was the primary element, then there wasn't really anything to worry about, at least from a health perspective.

Driven now out of sheer curiosity, we continued to watch them and found that there was a pattern to their disbursement. They worked in sections, and they never disbursed these trails over one area two days in a row, they were always moving from one section to another. The size of the area appeared to be about the length of the trails, about 50 miles by 50 miles.

We saw there was a direct correlation to an increased level of activity prior to incoming storms. We had also observed greater activity in the upper Midwest than other areas around the nation. Over the last few years, we have traveled across the nation, most recently, Houston. Although we did observe some of these trails, they were very few in numbers in comparison to the upper Midwest. Disbursement of small particles of dust into an atmosphere saturated with water vapor will create cirrus clouds, and this would explain the reason for both of these observations. The vast amounts of water found inland, specifically here in Minnesota, along with the water vapor in the air preceding storms would explain this increase in activity. The water vapor increases and maximizes its effects.

The other logical conclusion was, whoever was implementing this, also had loved ones who had to breathe it too. Even the person implementing it would not be exempt from the effects. Common sense here dictates that our government is doing what it can to implement a cooling effect, and for Minnesota and much of the eastern US, it seems to be working.

The world seems to be fixated on a desire to combat global warming. Current aerial distribution is operating with great vigor and reminds me of a military operation, efficient and precise in their operation, and also predictable. It is in this deep desire where we fail, as if we have the power to bring about change to our planet through forcing it to comply with our desires. This fault is found in man's ego and pride, but is only an illusion. If we try to fight against the nature of the planet, we will lose. The climates around the world can change, but only when we begin to understand that the world needs to be nurtured and not conquered. Mother Nature has been torn up, cut down, drained and dumped on, and today, she resembles a beaten woman. Mankind will never force her into submission and change her nature, it will be Mother Nature that will change mankind.

The difficulty one encounters when attempting to override system functions of a machine where the complete and entire operation of the system is not fully known, consequences will surely occur. These consequences are explored with this new understanding of our planet's heating cycle.

By placing an artificial canopy in the atmosphere, we will lower the amount of solar radiance upon the surface as it did in the '60s and '70s, but there are additional variables today that will alter the outcome. At the very least, by the laws of thermodynamics, it will slow the process. What is imperative is what we do at the surface level during this time so that such techniques can be discontinued. Not doing so will require this induction into the atmosphere to become a permanent part of our lives, and this is unacceptable. As soon as this ends, the heating will begin again due to the root cause still intact.

Weather modification is currently deployed, and the scientists that looked at this data concluded that this activity would decrease the earth's temperature. By doing this, they also support my data demonstrating that heat was rising due to the loss of the forests, jungles, and the prairies ground cover that once shaded the earth's surface. Air pollution once acted as an artificial canopy, applied by accident, that replaced this ground cover for a number of years. When the EPA cleared the skies, the heat returned once again from its initial cause, the solar energy being applied directly to the surface and the water runoff increasing this effect.

My research and observations indicate that the primary method used is through the deployment of particulates where water vapor is plentiful. Cirrus clouds quickly develop creating a haze effect upon the surface, depleting the amount of solar radiance. This in time will create a cooling effect upon the surface, but what can we expect and will there be additional impacts on our planet?

Based upon known physics, the answer is yes, but all of the side effects are not fully known. When we do not fully understand a machine and attempt to alter its operation, there will be side effects from our actions, and sometimes they can become catastrophic. The earth is no different and man's current responses to climate change reminds me of the man who placed a piece of cardboard in front of his radiator in the winter so that his car would generate more heat. This worked great until spring when his car overheated and left him stranded with a blown engine. He didn't fully understand the system operation, and if he had fully understood this and replaced his thermostat, he could have avoided this problem. Instead, he found a quick fix around a problem, implemented it, and paid the price later. This is what we are doing in our attempts to combat climate change.

Known Side Effects

Reducing the solar radiance upon the ground will decrease ground water evaporation and reduce precipitation over time. But this gain is being offset by the initiation of heat for oil and gas production and a continual decline in underground water. Minnesota experienced a decline in precipitation during the air pollution era by more than six inches per year due to the decline in surface temperatures, lowering the evaporation rate.

Our current situation is vastly different in comparison to the pollution era. Today, there is increased water in the atmosphere in order to create the haze effect through cirrus clouds. In the pollution era, this mainly consisted of low-level smog encompassing the ground. What we are doing today is relocating the water and this will create different effects. Additional particles in the atmosphere absorb and maintain the water vapor, allowing the atmosphere to act like a sponge. As this water level rises in the atmosphere, there will be a point of saturation. We have begun to see some of the consequences of this action in a rise in flash flooding as we have seen in cities like Chicago and Toronto over the past couple of years. These occurrences will rise in both frequency and levels and give way to late summer dry conditions.

The world is far different today than it was after WWII, and we need to consider our current heat gradients, the lower water levels, and the additional induction of heat caused from shale production to get an overall impact.

Carbon dioxide is taken in during a process called photosynthesis. This process only occurs during hours of sunlight. Reducing solar radiance upon the Earth will decrease the planet's ability to grow and sequester carbon dioxide. This will cause a marked rise in atmospheric carbon dioxide concentrations that will accumulate yearly. Additionally, this will cause a decline in vegetation growth affecting crop yields. Yield production relies heavily on proper water, nutrients, and solar radiance. By maintaining water and nutrient levels, but decreasing solar radiance, there will be a reduction in plant growth affecting crop yields. It can alter growing seasons and wreak havoc on farmers in their choice of crop selections, as it has the past couple of years.

Reducing the amount of solar radiance upon the surface will also reduce the solar energy potentially decreasing surface winds. This would adversely affect renewable energy sources such as wind turbines and solar energy.

Using these techniques can produce another side effect. The amount and degree of this effect is unknown. Adding elements into the atmosphere spreads out a storm out over a wider area and will weaken the front. By doing this, the storm is suppressed, and for society today, this is considered a positive result. This decreases the intensity of storms, reducing the level of damage that tornados, damaging winds, hail, and lightning strikes can cause. By doing this, there is also an unexpected consequence. Decreasing lightning also decreases the electrical field that surrounds the core. By doing this, the field declines in strength allowing more energy through, resulting in an additional heat factor. What may seem to be good for society today is not necessarily good for tomorrow.

Human behavior can alter due to the obscuring of sunlight and is known to lead to lethargy, depression, and other issues including suicide.

Because these particulates are airborne, the chemicals used may have side effects upon life on the planet, including humans. Inhalation effects of many chemicals often pose an increased risk with potential side effects and health issues, especially with the elderly, the weak, and the children.

Magnesium and calcium are two natural elements found in the soils. Increasing or decreasing either can cause acidic soils for those areas. Areas along the oceans have adapted for higher levels of magnesium found in the rain coming in from the oceans, while inland the vegetation has adapted for less. Increasing magnesium can be easily countered by the agricultural sector by increasing the calcium in its fertilizers, but this will not aid the natural vegetation. As a result, a variety of species may become strained.

Attempting to block the solar radiance upon the Earth without proper knowledge of the Earth's heating and cooling system is nothing more than a guess. Knowing it will offer cooling is one thing; knowing what you're doing with it is another. I know that science and our leaders are unaware that the only time spraying is needed is in the spring when the fields are barren and into the summer months because in the late fall, the fields are actually cooler than the forests and prairies. This is primarily due to the elimination of land cover and the reduction of the daytime sun. In the winter, it's useless and can only bring about side effects from an abundance of foreign materials. The primary focus needs to be centered upon the outer edges of a swollen atmosphere. This portion of the atmosphere needs to be compressed through aerial distribution and can be targeted. Instead, we are using a form of massive carpet bombing—throwing everything we have at it, hoping we will make a difference by creating a massive haze effect year around.

There are many feedback loops around the world such as the loss of the ice caps and glaciers that reflect 80% of the sun. When they turn into water, they absorb 80% of the sun, increasing their heat effects. In these feedbacks, there are two potential threats based upon heat and pressure that the United States needs to be concerned about.

First is the heating of the Rocky Mountains along the western edge of the North American Plate for oil shale. As this heat builds underground, it is carried out and away from the source and will follow the water basins. In Canada, these shale deposits follow along the Mackenzie River to the Arctic. This heat is forcing the permafrost up and out of the ground. Once this cold air is above ground, the air temperatures are warmer so this cold air resides along the surface and gathers over Eastern Canada and the Hudson Bay. This has become a pattern that has developed over the past few years and it has created some extreme temperature deviations resulting in massive temperature fluctuations. One half of our plate is being heated while the other half is being cooled. This will cause the cooled portion of the plate to contract while the other side of the plate expands. These different variables create stress cracks, and with our planet, this means earthquakes and potentially large volcanic eruptions. Yellowstone is not too far from this current activity and is being impacted by these processes.

On February 19, 2015, the following was reported: "Grizzlies waking up ahead of schedule in Yellowstone."* This heat is rising, but the cold in the east is overwhelming the news. Nature has instincts that we do not understand and are always a way of monitoring what is happening in their surroundings, and the bears waking a month ahead of normal and eating is not a good sign.

The second concern is what could arise out of the permafrost thawing. Deep within the ground, there are massive amounts of organic material that is currently thawing and the potential of some deadly virus surfacing after being dormant for tens of thousands of years rises. Even if there is no direct effect upon humanity, ripple effects will occur.

After reflecting upon all this research and data and attempting to find answers to our problems today, I came to realize that nature and man are in direct conflict with one another. Yes, we all say we love nature and wish to preserve it, but it also defies one of our basic core needs—cleanliness. This is noticed right away as the grass and foliage rises, so do our rodent populations. Standing water increases mosquitoes and some gophers are digging through your yard, flooding your foundation while the ants take over your kitchen. All of these problems in the world arise out

---

* http://www.foxnews.com/science/2015/02/19/grizzlies-waking-up-ahead-schedule-in-yellowstone/?ncid=AOL

of this conflict between nature and human activity. So the question becomes, how many people can the world support without adversely affecting nature?

On Sunday, February 22, 2015, the *Star Tribune* of the Twin Cities reported in their Science and Health Section, "Block the sun's rays to cool down Earth?" In this article, they explain testing using sea-salt particles to form clouds and cool the Earth as a possible solution to global warming. This is well beyond the testing stage, and it's only through pressure from the people that the government is now beginning to expose this process. It has become too big to hide.

THE EARTH: MAN'S ABILITY TO SUSTAIN

The Earth has a total landmass of 57.5 million square miles (36.8 billion acres). Of this, 50% is considered uninhabitable. Antarctica alone makes up 9.4% of this area; deserts, glaciers, and mountains make up the rest. Of this 50%, an estimated one half of this is unusable for farmland. The Tundra and Taiga make up 15.8% of this land alone with the majority located in Canada, Alaska, and Russia. This leaves us with an estimated 9.2 billion acres of potential land for farmland and today it is estimated that there are 5.1 billion acres available to be at 100%. Without expanding into Northern Canada or Siberia, today we can conclude that we are at a 45% usage rate for the world's population for farmland alone. Take into account all the homes, roads, malls, buildings, parking lots, and everything from treatment plants to sewage, we can consider the amount of arable land altered today is at 70%. In America, using calculations from the Department of Transportation, roads in America alone would pave the state of Kentucky.

Our early ancestors instituted the Homestead Act that allowed the government to control, monitor, and record the land for farm use. By World War I, 44% of the lower 48 states had been converted to farmland in order to support the war effort. This peaked at 59% in the 1950s and has since declined to make room for city and urban expansion. This does not include cities or urban development. When land capacity reaches maximum potential, farmland must decline as population grows. Since the 1950s, this has been occurring in the lower 48 states.

By the 1960s, the lower 48 states, along with Europe, Argentina, and Australia had no more land available to feed the ever growing world population. In response to this increasing need, land around the world was and continues to be aggressively converted in order to prevent food shortages, including jungles throughout the world. We began to see growth hormones and genetic engineering to increase yields and avoid jeopardizing our food resources. With 4.1 billion acres of farmland, we can assess that 0.58 acres per person is necessary today to sustain all of our food. Now we also need to include housing, roads, sanitation, work environments, oil, coal, electricity, etc. Today it is estimated that it requires 1.04 acres of cleared land to support the needs of every person in the world today. Looking at it from this perspective, we are highly efficient as a species.

At our current rate of use, we can conclude that when the population reaches 8.85 billion people, we will have exhausted all of our resources and completely decimated all the forests, prairies, and jungles by this time. After this point, all arable land will be exhausted, and we will need to move and learn to live and adapt to environments in areas such as Siberia, Yukon, and the Northwest Territories in order to feed the population. We will be at this point by 2027.

With 5.1 billion acres available, and 0.58 acres per person needed, the maximum potential population of the planet will reach its limits before it reaches 16 billion people based upon our current usage. This includes no new housing, stores, roads, etc., only farmland. At this point, all forests, prairies, and jungles will be gone and every tree and blade of grass wiped out in order to feed and maintain this population. This is an unrealistic number to attain, and an environmental collapse will occur well before this time, but gives us a total maximum. Science indicates that a 50% alteration will begin this collapse, and we are near this today.

Population Growth

Several years ago, I decided to look back in time and review my ancestors. I was able to track my heritage back to the 1800s before links began to breakdown. My mind began to question this because it was computer language. Beginning with me it was one, then two parents, four grandparents, eight great-grandparents, and so on. Understanding this,

I decided to go back to "year 0" and find out how many blood relatives I had when Jesus was on Earth and what the chances were of somehow being related to Mary. By using a figure of 25 years per generation, I had determined there was 78 generations between today and year 0. By using a calculator, I began punching it in, 1 times 2, times 2, times 2, for 78 times and came up with a number to be estimated at 151 septillion. This is 24 zeroes after your number or 151,000,000,000,000,000,000,000,000 people; we are all related to one another, and if you are between 25 and 50, it's twice this number. Here we experience the exponential power of doubling.

When I had finished my research on eruptions, I saw a pattern in time frames. I had one question on my mind, what were our ancestors doing back then? There is a cycle within mankind, causing disruptions in the planet's climate, and today is not the first time we have experienced this, but the first time in known history that we have been able to attain such a large population simultaneously at one time around the world. By implementing the power of numbers, we can see how this repetitive cycle occurs. All of the information, I have acquired is useless if we do not look at how we can alter our future and implement changes. I needed to answer the question people ask me at this point, "Can we alter our course?" The following figures are simple math equations and any future alterations to this equation will require all of humanity to come together as a whole.

I have heard from many people over the years that the world has cycles. Oftentimes, when discussing global warming, they refer to a time when Greenland was occupied during the European warming and expansion era that occurred between ad 1000 and 1250. Throughout this research, I continued to find time spans that ranged from 200 to 250 years that appeared to be a cycle. This appeared in North America since the time of the Mayans and continually repeated itself. But there were two known events that prompted me to look deeper into this cycle as a planetary response to man and not a normal cycle of the planet. These two events were the European warming and expansion from 1000 to 1250 and the Black Plague in mid-1300s followed by the migration to America in the 1600s, a totally different plate.

If there are 1,000 people, 500 couples, and they all have one child each (two per couple), the population of this community will grow to 3,000 people considering three generations take place before one generation replaces another. At this point, the population will stabilize as each generation replaces the older generation. With one acre per person, the amount of land needed to support the population of this community will be sustained at less than five square miles and will not need to grow beyond this.

Years	Begin	Births	Deaths	Total Population
0–25	1,000	1,000`	0	2,000
25–49	2,000	1,000	0	3,000
50–74	3,000	1,000	-1,000	3,000
75–99	3,000	1,000	-1,000	3,000
100–124	3,000	1,000	-1,000	3,000

If you take the same figures but increase the number of children to two per person (4 per couples), the power of doubling reveals the results.

0–25	1,000	2,000	0	3,000
25–49	3,000	4,000	0	7,000
50–74	7,000	8,000	-1,000	14,000
75–99	14,000	16,000	-2,000	28,000
100–124	28,000	32,000	-4,000	56,000
125–149	56,000	64,000	-8,000	112,000
150–174	112,000	128,000	-16,000	224,000
175–199	224,000	256,000	-32,000	448,000
200–224	448,000	512,000	-64,000	896,000
225–249	896,000	1,024,000	-128,000	1,792,000

Because each person requires one acre of land to sustain their life, it is now necessary to increase the land necessary to feed the expanding population from 3,000 acres to over 1.8 million acres in nine generations. The amount of land needed to sustain this level of population would rise from less than five square miles to 2,912 square miles, larger than the state of Delaware. There are no problems as long as there are no neighbors, but when neighbors are present, problems begin as growth becomes restrained and resources dwindle away. This is the power of doubling and explains our current struggles as we near this precipice.

Based upon these multiples, if humanity began with only one woman and one man and we sustained a population growth of four children per couple, within 29 generations there would be 2.147 billion births and by the thirtieth generation collapse will occur, and this happens in only 725 years. With three children per couple, we are looking at less than 1,000 years.

With a collapse occurring every 10 generations, 250 years, all of these years seem like a very long time ago when reflecting upon our life span. Today, this is looking at the year 1764, a very long time ago in human years. We need to understand that the average person will live with, know, and love five generations of ancestors; so the question becomes, where are we? If we are at the beginning or middle, there is great responsibility; if we are near the end, there is preparation that must begin, and we need to pass on information and knowledge to our children so that they can avoid this cycle that humanity develops and continually brings about great pain and suffering upon himself.

Now we will take these same figures and have each couple raise only one child each and now we see a decline in population. The population will begin to decline but takes time just as it does to grow.

Year	Begin	Births	Deaths	Total Population
0–25	1,000	500	0	1,500
25–50	1,500	250	0	1,750
50–75	1,750	113	-1,000	863
75–100	863	57	-500	420
100–125	420	29	-250	199

Current Population

The United Nations has been working upon the accuracy of the world population for years and the seven billion mark was celebrated on Oct 31, 2011. Using the data on the world's population we can see our change over time.

Year	Time Frame	#(in billions)	% change in total	% change per year
1804		1		
1927	123 yrs	2	100	0.81
1960	33 yrs	3	50	1.52
1974	14 yrs	4	33	2.36
1987	13 yrs	5	25	1.92
1999	12 yrs	6	20	1.67
2011	12 yrs	7	17	1.42
2023	***12 yrs***	***8***	***15***	***1.25***

Numbers in bold italic are estimated projections based upon a hopeful reduction of birth rates worldwide. We grew from six billion to seven billion people in only 12 years, 4,380 days. At this rate, we find our growth rate, births minus deaths, averaging 228,310 additional people every day.

Three hypothetical population growths will be used to achieve future population growth scenarios with no famines, plagues, or massive worldwide catastrophes. We need to recall that the tsunami that killed so many people in Sumatra set our population growth back just over one day. The tsunami in Japan set us back only a few hours.

Using three variables, the current rate of 1.42% that is the most recent, a reduced rate of 1.2%, and the Chinese strict implementation of population control that has resulted in a rise in their population growth of 1.07%. We also need to take into consideration that this population growth would need to have been implemented and adhered to since October 31, 2011.

# (in billions)	@1.07% per yr	@1.2% per yr	@1.42% per yr
8	2024	2022	2021
9	2036	2033	2030
10	2046	2042	2038
11	2055	2050	2045
12	2064	2058	2051
13	2071	2065	2057
14	2079	2071	2063
15	2085	2077	2068
16	2091	2082	2073

The US Census Bureau indicates a world population of nine billion will be reached in the year 2046. In order for this to occur, they anticipate the population to rise 12.9% (seven to nine billion) over 35 years (2011–2046). This is a population growth of only 0.368% per year, less than half the 1804–1927 era. Reality indicates that even with strict laws regarding population growth like China, population of the world will be at 10 billion in 2046. As the population rises, the rate of rise to the next billion quickens.

Our population growth is not a result of planned pregnancies, but a result of unplanned pregnancies resulting from sexual activity. As a result, this low rate will not be attained through a reduction in birth rates, but by wars, famines, plagues, or other catastrophes that would result in surpassing the planet's ability to supply the necessary resources. The United Nations indicates the same kind of statistics and a population rise of only 0.33% per year after 2050. These figures indicate a slow rise in deaths of at least one billion people by 2046, or 33.3 million deaths from unnatural causes per year between now and 2046.

Population growth greatly increased with advancements in medicine and medical practices in the early 1800s. Further technologies such as penicillin, antibiotics, immunizations, and electricity have all contributed to a very low infant and child mortality rate today, even when compared to only 50 years ago. Current technology and advances in medicine have allowed us to increase the longevity and health of the people bringing in another extended generation. Comparatively, the impact on the longevity and quality of life is shadowed in comparison to the impact of child and infant mortality on the world population. Along with this technology, we have forgotten that there is also a responsibility in preventing overpopulation and potential self-destruction.

In order to hit a zero population growth, which is necessary for civilization to continue, there must be at least 228,310 less births or increased deaths per day from unnatural causes. Although we cannot control the world population, regions that conform to these alterations and enact laws to prevent any further environmental degradation will sustain best. It's for this reason that action must take place no matter how small it may seem.

In a democratic environment, it is the responsibility of its leaders to create awareness and enact laws to protect its citizens, especially if the threat to its people are due to the activities of the people themselves. This allows the people to become aware and enact measures in their own lives to promote change. This has to come about willingly by the people, and there is nothing any one person can do. It's a culmination of what everyone must do through public awareness and distributed knowledge that we must all do together.

Historical Population

The last major impact upon the population of the planet was the black plague in the mid-1300s. Even the wars since this time do not come close to the percentage of lives lost during this brief period of time in our history. I am sure the people who lived through these times thought it was the end of days coming upon them, but soon it passed and life went on again. The population of the world was decimated, and it began growing slowly again after this event.

As the population grew after the plague, North America was found and vast amounts of resources were opened up, allowing for vast population growth without any major crisis. Due to technological advances and increased farming production and yields, we have been able to grow and prosper without the threat of catastrophic worldwide collapse. Even with this as our advantage, there are still limits to our planet. At some point, the population will naturally overrun its availability to produce the resources needed to support its numbers. When this occurs, history will repeat itself through wars, chaos, and anarchy.

Prior to the 1960s, birth control was not a part of our lives. Understanding this, we can begin to understand previous civilizations and their potential for areas and size of their populations. In these estimates of growth, we can conclude that a starting population of one million people would exceed one billion within 250 years at a rate of four children per couple.

Through the use of fossil fuels promoting heating and cooling, we are now able to sustain in environments once uninhabitable. Although areas in the Midwest still prosper, without the availability of air-conditioning, mankind would be migrating. Because man is able to sustain this environment, we need to understand that nature itself cannot withstand these variations and will migrate. Population has continued to grow and prosper through the years and now, 213 years later, we see sexual promiscuity and population growth that is out of control.

Many people claim that we are reducing in population, but this is just an illusion by one's own observance in their own community and does not account for the world. Today, we are a world society and one end of the world affects other regions of the world. Some think technology will be able to overcome many of these obstacles, but they are filled with illusions of power and our ability to control the Earth. The Earth is nothing to be controlled, it's to be nurtured and cared for and wants to be left alone, but we can help her now that we understand what she needs.

It is very speculative as to the exact dates the world reaches specific populations, but the information age has made the accuracy far greater over the years. The exact date when the world reached six billion people tends to range in dates from June to October of 1999. For reaching seven billion people, the dates range from March of 2012 until October 31, 2012, when the United Nations celebrated this milestone. Based upon these values, we can determine that the population growth rate averaged between 205,549 to 220,750 people added per day from 1999 to 2012.

Today, December 6, 2014, they state there are 7,279,375,317 people in the world today. Within a few months, we will have added a population equivalent to the United States' total population in just a few short years. To remain conservative, we will use the earlier date of March of 2012 to achieve the lowest figure. This is 970 days and 279,375,317 people added, making an average of 288,015 people being added to our world every day today. Clearly increased over the previous era, but remaining stable when we consider the population increase will rise this rate. At this rate, we will be at eight billion people no later than 2,502 days from today, or 6.85 years. This comes to a date no later than November of 2021 at our current rate.

The total amount of arable land available in the world for agricultural development to feed and house the people of the world can be estimated at 16.9 billion acres. Currently, 44% of this has been altered for the purpose of sustaining life for the seven billion people. This comes to 7.43 billion acres, 1.04 acres per person to sustain all our needs throughout the year. This is 209 square feet per person today. The overall percentage of population has been declining since the mid-1900s, but due to the increased population, there continues to be an average of 228,310 people added per day to the planet when you add the births and subtract the deaths. At our current rate of population growth and use, it will be necessary to alter more than 237,442 acres a day to maintain our way of life. As some land alters, like the Gobi Desert, growing three thousand square kilometers per year, a more aggressive approach to deforestation will be necessary to sustain the world's need for food beyond this amount.

Mathematically, it must come to an end. The question is no longer if, but when the world becomes a place of chaos. We won't run out of water or oil, our shortage will come through increased population, placing an ever increasing demand on food resources. This competition for food will not be limited to mankind, but throughout all of nature as every species attempts to survive.

As populations increase and the densities rise, the potential for an outbreak similar to the black plague increases proportionately with it. Ebola wasn't an issue when I first wrote this, but demonstrates the potential threat that awaits mankind. We are coming to a point that men throughout history have discussed, debated, talked, and written about for hundreds of years. Conflicts and wars over land and resources are inevitable.

It would be very realistic to see the population of the world go from eight to six billion quickly, a one in four ratio. If this was to occur over a period of ten years, we can calculate the repercussions of this event now. There would be 547,945 deaths per day when this time comes, the holocaust occurring every eleven days for ten years. If the time frame is less, the death toll will be more. Because there would be continued births during this time, we would also have to consider their lives into our calculations that, at our current rate, would put more than 800 million more lives on the Earth.

Population Awareness

Looking at it from the perspective that if every two people have two children and they have two children, we would maintain population stability indefinitely. Of course, not everyone reproduces, so some level above two is necessary, which is dependent upon population densities. Population growth has no effect at this level and anything more adds to the overall growth of a society exponentially as each generation passes. History shows that when a society progresses beyond its tenth generation, collapse occurs due to a lack of resources. It's mathematical and something we should all have learned in school, home, and religious upbringing, so the question becomes, why not?

Keeping people from knowing and understanding exponential growth means increased power and control. In the past, it has meant the difference between a nation, preventing itself from being overrun or perhaps a religion flourishing so they promote growth through population expansion. This brings with it more soldiers, invaders, defenders, and revenue for a leader. If you collect one dollar from 50 people, there is only 50 dollars, but if you collect one dollar from 300 million people, you have 300 million dollars and the impact upon the people remains the same. In our world today, much of this is what provides economic power and control for both individuals and nations. Whatever the reason, power and control is at the core.

Many do not want to be told they must control the amount of children they have in their life. A man may want his genetic heritage to continue by having multiple children or perhaps engage in multiple relationships. In his actions, he disregards the lives of his own descendents and exempts himself of any responsibility his actions will have on them. At the core, his desire drives him, and here we find both power and control wrapped up within his own ego.

For a woman, she wants to have the freedom to choose in regard to her body, and I fully agree with this. No one should ever have any power or control over her, and I understand and respect this. Although she may not want to hear that she needs to control the number of children she has, she needs to know and fully understand the impacts of her choices, apply them to her life, and then teach this to her children. In this decision to ignore this knowledge, we find a desire to control, and the illusion of being self-empowered.

Governments and big business will not agree with population reduction, as this would mean a reduction in growth. Less people means less revenue for governments. Declining population means fewer resources needed, resulting in less fastfood services and retailers, less medical needs, less everything. This contradicts Wall Street's desire for growth, so money and power will naturally oppose population reduction.

I believe it is wrong to establish laws that require people to conform to population control or any form of procedure that would leave a person unable to give birth. Yet, it is equally wrong to bring up our children in ignorance knowing the truth and keeping this from them as if this is in their best interest. What is in the world's best interest is to know the truth, good or bad, and to pass on this basic knowledge of the Earth to our children. Understanding population growth is the key to the longevity of a society.

Countries, regions, and communities that make the necessary alterations will endure the best, while other areas of the world that neglect this warning will be subject to massive struggles and no one will be exempt. Because we start these population figures with only healthy adults of childrearing age, the impacts would begin immediately. Within 50 years, we could see our population down to less than six billion people, but would require two generations, 50 years of a reproduction rate of one woman, one child. Without addressing this first and foremost, any attempt to alter our outcome will be in vain.

Everyone will experience five generations during the course of a typical lifetime, from grandparents to grandchildren. Within 10 generations, societies collapse so the question becomes, where are we in the cycle?

Understand Our Past

In the 1930s, 2.5 million people were displaced due to the heat from the great Dust Bowl. According the Census Bureau of 1930, this comes to 22% of the population for New Mexico, Texas, Oklahoma, Kansas, and Colorado. Today, 22% of this area's population equals 8.47 million people.

In order to better understand our past, I believe we have to imagine one day being placed on a remote island called "Earth." If we consider that they had processing skills as we have today, it wouldn't have taken them long to discover many things, like the day and night routine, the change of seasons, the alterations of the moon and stars, they just lacked the names or science behind it. Within the first year, we would be designing some form of a calendar. We would want to prepare for the alterations and changes in seasons because the cold weather nearly killed you last year. By looking up to the sun, moon, and stars, we would have created a calendar early on. Soon, you could navigate at night by watchful observations of the night sky. This would aid in planting, harvesting, and preparing for the winter months. You found apples, oranges, and various fruits and vegetables, and soon you found that where you disposed of the seeds, plants grew. Today, we find many past civilizations having structures aligned according to seasonal changes, and this would be expected.

Today, the clock is a given part of society, but growing up, it wasn't this simple. Everything ran according to the clock: work, school, banks, doctor's appointments, etc. Around the house, especially in my father and grandfather's era, this was even more important and was vital for the family. If we ever needed to calibrate our times, I always remember looking to Dad or Grandpa for having the right time on their watch.

It is important for us to look to our past and better understand our history. It is our nature to look at past generations as being behind the times and not up-to-date. By doing this, we exalt ourselves and our generation as more advanced than the one that preceded it. This easily allows our thoughts and imagination to think back in time to our ancestors as being even less advanced—farmers, sheep herders, hunter and gatherers, stone age, or barbaric perhaps. It is time for us to let go of these thoughts and begin to realize that our ancestors were as intelligent as we are today, and in many cases, more so.

I think back to people such as Aristotle, Newton, DaVinci, Einstein, and many other historical figures who, if part of the world today, would be redesigning the way we look at our world.

Life would naturally be driven by both the daily and lunar cycles. As I think about this, leap year is something that would require multiple generations and a documentation of knowledge passed on from one generation to the next. Over a period of 60 years, the clock would alter by only fifteen days.

When I traveled to Egypt, I saw many items that amazed me. Engineering marvels to folding beds, I was astonished at the workmanship. Then I saw a ceramic jar with an iron core wrapped with a copper wire. When fruit juice, which is an acid, was poured into the container, voltage was created. A battery was in use thousands of years before we recreated this again.

By understanding population growth and historical timelines, we can begin to understand how large the populations were during the Mayan, Anasazi, Inca, and Aztec empires. A vast amount of land alteration by a very large population is needed for these large scale volcanic eruptions to have occurred. We can see that Atlantis and other marvels are not only probable, but would be a natural form of progression and destruction within our history.

Man has repeatedly created societies that only collapse as far back as our known history takes us. By understanding and sharing this simple math, we can begin to live within the confines of this planet. It's not a planet that we can conquer and sell, but to live with and nurture. This can only be accomplished by following the rules and restrictions of this world.

Here, we find the cause of global warming and climate change—it is within the population growth and densities of our societies. It's not an external fight we must overcome, but an internal one. Once overcome, humanity will be able to move to a new level that surpasses our wildest imaginations. Humanity needs to change its direction from battling global warming to an understanding of it, and then implement measures to offset these changes. We must first start by learning to live within the constraints of our environment. The Earth is finite, but our ability to populate is infinite.

We cannot save the world, but if our actions and laws today can save the life of even one child in the future, then any and all laws implemented would be worthy of being addressed, and every action that we can assert today should be applied. If we decide to do nothing, there are consequences that come with this choice that will have even greater impacts on the people. This is a basic two-step process with many additional issues that needs to be implemented along the way. Reverse population growth through voluntary implementation and replanting and protection of the Earth's resources.

The wealth of Wall Street will be needed because their wealth was built upon the resources of this world. This is the first stumbling block we will encounter and it's here, in the money and power. It is because of this that we will continue to argue and discuss the world's problems and the potential options and solutions without initiating any appropriate action until it becomes too late. We had no advanced warning when the stock market fell, and we will also have no advanced warning of food shortages or plagues either. Such an issue, when it comes to pass, will destroy the economy and create overnight civil unrest. We rely on the Earth providing ample rain and sunshine throughout a season to provide the food necessary, and here we are playing with fire.

When confronted with the following information, people ask and wonder, what can I do about it? There are many things that can be done, but without the proper knowledge, nothing will be done. Even with understanding, the sacrifices and alterations in our daily lives necessary will be far too great for us to change, and change only comes about when we meet a precipice, and at that time, it will be too late.

Change

To understand the impact of any one item, like the clearing of one acre of land, think of it in the extreme to understand its effects when calculating the impact of seven billion people. Collecting just one penny from everyone on the planet would bring in seven million dollars in revenue, and if you and everyone else planted just one tree, we would have planted more than seven billion. No single person put us into our predicament, and no one person will pull us out of it. The decisions and choices we make today will determine the destiny for so many of our children, and a decision to do nothing will only increase the suffering.

The key to change is found in the understanding, sharing, and teaching of the mathematical formula for population growth. When we are born, we become aware of our surroundings, and we have many experiences that increase our knowledge as we mature. Nowhere are we taught that a woman should refrain from having more than two children, and yet this is a simple mathematical equation that needs to be taught as a way of life from birth. This includes home, school, math, sex education, and religious upbringing. This should be common knowledge among all mankind. A woman would refrain from having too many children if she truly knew that in the years to come that her descendents would need to endure intense suffering as a result of her choices today.

The other topic that needs to be addressed is sterilization. Because this is easier for men and is a far less invasive surgery, men need to carefully consider having this procedure if they are sexually active, especially if they already have children. Today, this procedure needs to be available to all men and women at no cost, and there needs to be an appropriate level of privacy too.

Next, it is imperative that wherever in-ground heating is occurring, these processes be terminated immediately. This traps the heat in the ground where it builds continuously without release and produces a very high level of instability. These practices are not limited to volumetric heating or oil shale production.

All attempts to manipulate the weather must stop immediately and allow the planet to react naturally. Attempting to manipulate the weather patterns is resulting in many side effects and what will aid the planet the most is for our activity to be the least. This is also known as playing possum. The Earth has the appropriate mechanisms in place for self-correction, and we need to allow it to adjust for the changes we have made.

Mankind has altered a disproportionate amount of the land surface of the planet that inevitably has caused a rapid alteration of the surface climates and deterioration of the planet's ecosystems. Our planet must be regenerated, and we can help nurture her back to health. Doing so will offer the best security for future generations. In order to accomplish this, there is a need for manpower, and we have a great asset in our schools. An estimated 17% of our population resides within this school age group. This is a workforce of an estimated 50 million children and teenagers in school during the spring. If every student was to germinate 10 local trees and plant them within their communities we could germinate 500 million trees a year in the United States alone.

There is a workforce of nearly 150 million people in the United States and millions of homeowners. The accumulation of this work force is imperative to aid in implementing these changes. Cities and local communities will need to implement plans of replanting, preparing the land, and maintaining care for these trees that they will plant to assure their survival. After rooted, natural vegetation doesn't need any continuing care and will be self-supporting.

Cities and states will need to prepare land for the purposes of planting. Empty paved lots can be torn up and replanted. Power lines should be buried rather than trimming the trees and trees can take the place of all the telephone poles. Everyone can take a portion of their land and increase the natural land cover for their region. Roads and freeways need to be lined with trees, shading the surface and many roads, intersections, and parking facilities can be moved underground. Layered parking and retail space can offer the same volume in a smaller area. Understanding that there is a need for deep-rooted vegetation, we can begin to implement many changes.

While down in Louisiana after hurricane Katrina, I began to wonder why the logging industry wasn't taking advantage of the massive amount of lumber available after this huge storm. While naturalists attempted to leave the fallen lumber in place as part of the natural ecosystem, we decided to import lumber and chop it down from other areas of the nation to rebuild theirs. From a natural perspective, the lumber industry needs to leave the forest alone and become storm chasers using lumber from storm damage to supply these necessary resources.

Water is a vital part of our planet's cooling system. In order to reduce the levels of underground water extraction, it would be necessary to retain the water in the spring to use for farming and urban use. In order to hold back this volume of water, much of the landscape would need to be altered that will create rising issues, and deep discussions over areas that would be flooded. In addition, farming practices and crop selections need to alter and our farming techniques would need to change. Utilizing wet crops such as rice offers far more cooling and carbon dioxide sequestering than dry farming due to the cooling factor that the water provides. Additionally, this water promotes life and growth, adding more carbon dioxide sequestration than dry crops. Such farming techniques, specifically in South East Asia, have provided widespread population growth over the years with a decreased speed of impact.

Crops such as apples, olives, cherry orchards, or perhaps raspberry fields are far more productive for the Earth and allow for both cooling and food production while maintaining stability in a natural environment.

There are countless ways we can help the Earth, just as there are also countless ways we can continue to destroy it. These are just a few ideas and what I hope for is that together we can build upon them. I can only hope that this will bring about a vast amount of discussions around the world and demonstrate how everyone can play a role in destroying or paving the future for our children.

Conclusion

Technology today is doubling every two years and soon it will become every two months. As our knowledge expands, our wisdom seems to be falling behind. Although we may be able to store and process massive amounts of data, what good is all this information if it can't be applied to anything more than finding out what kind of shoes Mary would like? Innovation is what our world is in need of today and in order to accomplish this, we need to expel many concepts and ideas that have been engrained into us. Throughout our history, mankind has been able to use his ingenuity and imagination to be able to overcome many obstacles, and we need to generate this back into the world today.

While overseas a couple of years ago, I met a number of college students that were researching solar power, and we were discussing the potential for this energy. I brought up the subject of creating an electric car that could potentially regenerate and propel itself. They looked at me and laughed and told me that such a thing could never happen. I listened to them go on for several minutes about the physics and theories that were involved. I then asked them if they were sure that something of this nature could never be done, and they reassured me that it was not possible.

Inventive thinking requires thinking outside of the box and overcoming obstacles through innovative ideas. They were limiting their thinking within electrical and perpetual motion. With a smile on my face, I looked at them and told them, "Your great-grandfathers were able to think outside the box and overcome this obstacle many years ago." I then removed my watch and told them, "This is a self-winding watch, and it has no battery and never needs to be wound, and my grandfather who gave it to me told me that as long as I live it will work, and so far he is right."

They were silent and I then began to discuss ideas that I had that could be implemented that could potentially create a car that would generate enough power to regenerate itself through other forces at work and using current technology. Soon, they were all jumping in discussing ideas and concepts. As I left, I turned back and saw them all discussing among themselves, and I walked away with a big smile on my face. If there is one thing this book does, I hope it does for you what it did for them—to make you think.

As I come to a close, I have had years to research this material, many of you will have had only days to read it. These findings and conclusions are based upon known physics and mathematical facts of our world from researching our planet using basic understanding of heating and cooling, and many additional materials found along the way.

For myself, I have had to go through many transformations over the years. There is a typical course one endures when they realize their life is coming to an end, and this book emphasizes that change will be our only alternative and we have no choice in this, it is just a question of when.

As I reflect over these past several years, I have realized that I have struggled through many of the feelings and emotions that I experienced with the death of my mother, the many stages of grief. In the beginning, I was filled with denial and isolation as I indulged into this work. I wanted to find an alternative, and I kept double and triple checking my figures, but they were accurate.

When all the calculations kept coming back to the same conclusions, I decided to try to do something about it. My anger over our situation and our loss of direction in combating climate change compelled me to reach out to politicians and institutions in an attempt to invoke change, but no one wanted to listen.

This brought about frustration, and I became depressed over our situation. But every time I saw my grandchildren, I continued to remain hopeful in some way. I began to accept the fact that we will endure a period of time that is approaching quickly, and it will be filled with much suffering. Just like a tsunami wave off in a distance or an avalanche on its way down a mountainside, it's on its way and now we can all see it, but no one can truly predict when it will hit. The longer it takes, the greater the impact it will have due to the constant rise in population levels and densities.

Acceptance offers not only freedom but hope, and this is what I wish to share with you. Acceptance may seem as if one is giving up, but it's the polar opposite. It's a release from the self and one's own desires, and a passionate longing for the betterment of others. It's a realization that our lives will come to an end, and realizing this prematurely in life does create a life of peace in a world surrounded by chaos.

Acceptance means prioritizing our life and realizing that material items can be replaced, debts can be paid, and money can be made, but time can never return to us. We realize that family and our relationship with others is a higher priority, and we make this an important part of our life. Our car breaks down or we encounter a medical issue that we struggle through. Perhaps the death of a loved one becomes transient because we know that this fate awaits us too. Material items are meaningless and money becomes only a tool of trade with no other worth. We become content with what we have and lose the desires for the self. We may become frustrated at times, and we can still become angry, but its direction and purpose are understood and remain focused on the larger picture, our future and this is in the children.

When collapse comes, the rich man will lose everything, but the homeless man will awaken and have lost nothing. And if this homeless man has hope, he has all the wealth a man will need.

We need to do everything we can today to prepare the way for these children, both the ones with us today and the ones that are yet to come. If you do not have a child, then consider just one. The more the mouths one must feed, the more difficult life will be when severe hardships begin. Do not live in a state of fear, but a life of direction with purpose and help others understand this knowledge. Do not bring up the children in this state of fear, allow them to be children and prepare their future for them. It is up to everyone to accept this situation at your own pace, but it is nothing to fear or worry about, but something we do need to prepare ourselves for. There is no faster way to collapse capitalism the way we know it today than to run out of food to feed the people and initiate food rationing. This is no longer a question of if, but when. If it weren't for genetic engineering and growth hormones to increase farm yields today, we would already be at this precipice.

The data and information accumulated from this independent research should be able to help aid science greatly in their further research and discovery of climate changes. Using my years of troubleshooting and bringing this field into the scientific world has brought me down unique paths of exploration, and this is because of the diverse knowledge that the science field has given to me.

In the future, the children of today may hopefully read this as adults, and if they do, I have something I want them to know. We messed up your planet and for this I ask for your forgiveness because we were young, ambitious, and creative, yet we were also ignorant to our planet and felt it was something we could consume and own. Our greed and ego engulfed us and we looked at the world as a source of profit and personal gain, and we devoured her, stripping her of necessary resources to satisfy our own desires. We developed a society that engulfed us from birth into this world—a world that catered to a select few. The governments were bought, and they became lost in their directions and were driven by greed. For all of this, I must ask for your forgiveness and to learn from our mistakes and to prevent this from happening to you. But mostly, respect and care for this planet, it's the only one you have.

So now as I come to a close and sit back and reflect upon our planet, the rotation, speed, axial tilt, orbit, moon, and sun, how they all work in beautiful harmony with one another. All of it reminds me of a finely tuned machine and it's here, in this understanding that has allowed me to see beyond what can't be seen. As I look closer to the water, atmosphere, vegetation, oceans, and then the human DNA, I begin to see so much more. Then I look into the eyes of my children and my grandchildren and I see what can't be seen—true love. Today, there has become so much tension between people over the issue of creation versus evolution, science versus religion. All of this data and information is only the beginning of understanding our planet, and there is so much that I do not know. What I do know after spending countless hours researching this planet of ours that we call our home is that this world was not something that occurred through a series of events, circumstances and consequences, but this world of ours was created to evolve.

NOTES

Searching for the Source

1. https://www.youtube.com/watch?v=1Oteyq4QywU&list=FLHfhQHRx8w0yH9XJcAMn4eg
2. http://energy.gov/energysaver/articles/landscaping-shade
3. http://www.epa.gov/heatisland/mitigation/trees.htm
4. http://ccc.atmos.colostate.edu/cgi-bin/mlydb.pl

Eruption Research "Pressure"

1. http://www.geography.learnontheinternet.co.uk/topics/structureofearth.html
2. http://www.nationmaster.com/graph/agr_agr_lan_of_lan_are-agriculture-agricultural-land-of-area
3. http://en.worldstat.info/North_America/Canada/Land
4. http://en.wikipedia.org/wiki/Geography_of_Russia
5. http://www.nationmaster.com/graph/agr_agr_lan_of_lan_are-agriculture-agricultural-land-of-area

6. http://www.nationmaster.com/graph/agr_agr_lan_of_lan_are-agriculture-agricultural-land-of-area
7. http://www.abs.gov.au/ausstats/abs@.nsf/Products/7121.0~2010-11~Main+Features~Land+Use?OpenDocument
8. http://ec.europa.eu/agriculture/envir/report/en/terr_en/report.htm
9. http://afe.easia.columbia.edu/special/china_1950_population.htm
10. http://inspectapedia.com/exterior/Coefficients_of_Expansion.htm
11. http://data.giss.nasa.gov/gistemp/graphs_v3/

Land Alterations and Their Effects

1. http://www.grida.no/publications/other/ipcc_sr/?src=/climate/ipcc/land_use/019.htm
2. http://cdiac.ornl.gov/trends/co2/lawdome.html
3. http://www.usatoday.com/story/weather/2014/01/15/weather-climate-report-2013/4490871/
4. http://www.seagrant.umn.edu/superior/facts
5. http://www.waterlaws.com/venture/calhoun.html
6. http://forestry.about.com/od/silviculture/a/Estimating-A-Trees-Age.htm

7. http://my.extension.uiuc.edu/documents/1722110809110911/Nebraska%20producing,%20harvesting%20and%20processing%20firewood.pdf
8. http://www.fs.usda.gov/Internet/FSE_DOCUMENTS/stelprdb5269813.pdf
9. http://www.fs.fed.us/ne/warren/longterm.htm
10. http://ga.water.usgs.gov/edu/wuir.html
11. http://ga.water.usgs.gov/edu/wuir.html
12. http://www.epa.gov/agriculture/ag101/landuse.html
13. http://www.epa.gov/oecaagct/forestry.html
14. http://www.epa.gov/agriculture/ag101/landuse.html

Heat Gradients and Oceanic Circulation

1. http://ase.tufts.edu/cosmos/print_images.asp?id=38
2. http://exploration.grc.nasa.gov/education/rocket/moon.html
3. Burnham, Alan K. (2003-08-20) (PDF). *Slow Radio-Frequency Processing of Large Oil Shale Volumes to Produce Petroleum-like Shale Oil.* Lawrence Livermore National Laboratory. UCRL-ID-155045. https://e-reports-ext.llnl.gov/pdf/243505.pdf. Retrieved 2007-06-28.
4. http://www.nasa.gov/topics/earth/features/2012-poleReversal.html

5. http://blogs.scientificamerican.com/observations/2013/05/09/400-ppm-carbon-dioxide-in-the-atmosphere-reaches-prehistoric-levels/
6. http://csat.au.af.mil/2025/volume3/vol3ch15.pdf
7. http://www-pm.larc.nasa.gov/sass/pub/journals/Duda.JAS.03.pdf